The Intergovernmental Platform on Biodiversity and Ecosystem Services (IPBES)

Twenty years after the Convention on Biological Diversity (CBD) entered into force, the founding of the Intergovernmental Platform on Biodiversity and Ecosystem Services (IPBES) in 2012 was the outcome of a long process of setting biodiversity issues at the top of the global environmental agenda. With contributions from more than a dozen renowned researchers in political science, law and sociology, this book analyzes IPBES functioning and challenges in terms of the knowledge selection process and actors involved.

The book reveals that, through its conceptual framework, IPBES promotes a pluralistic view of nature that calls for a broadening of the disciplinary frontiers. It combines natural science and social science research and also includes indigenous and local knowledge. IPBES is considered to represent the institutionalization of a permanent knowledge assessment on biodiversity and is often referred to as an IPCC success story, constituting a new stage in global environmental governance. In analyzing the knowledge selection process for IPBES decision making, the book better situates IPBES within the biodiversity and global governance domain. It ultimately argues that the establishment of IPBES provides a new opportunity to coordinate the different international conventions (CBD, RAMSAR, CITES, etc.) and initiatives (international assessment of marine biology, scientific programs, funding, etc.).

Marie Hrabanski and **Denis Pesche** are researchers in political sociology at CIRAD (Centre International de Recherche Agronomique pour le Développement – French Agricultural Research Centre for International Development), Montpellier, France. Within the research unit UMR ART-Dev, their work focuses on international biodiversity policies.

Routledge Studies in Biodiversity Politics and Management

The Intergovernmental Platform on Biodiversity and Ecosystem Services (IPBES)

Meeting the challenges of biodiversity conservation and governance

Edited by
Marie Hrabanski and Denis Pesche

Taylor & Francis Group

LONDON AND NEW YORK

from Routledge

First published 2017 by Routledge

2 Park Square, Milton Park, Abingdon, Oxfordshire OX14 4RN
711 Third Avenue, New York, NY 10017

Routledge is an imprint of the Taylor & Francis Group, an informa business

First issued in paperback 2018

British Library Cataloguing in Publication Data
A catalogue record for this book is available from the British Library

Library of Congress Cataloging in Publication Data
Names: Hrabanski, Marie, 1980- editor. | Pesche, Denis, editor.
Title: The Intergovernmental Platform on Biodiversity and Ecosystem Services (IPBES) : meeting the challenges of biodiversity conservation and governance / edited by Marie Hrabanski and Denis Pesche.
Description: London : Routledge, 2016. | Includes bibliographical references and index.
Identifiers: LCCN 2016017841| ISBN 9781138121256 (hbk) | ISBN 9781315651095 (pbk)
Subjects: LCSH: Intergovernmental Platform on Biodiversity and Ecosystem Services. | Biodiversity--International cooperation.
Classification: LCC QH541.15.B56 I577 2016 | DDC 333.95/16--dc23
LC record available at https://lccn.loc.gov/2016017841

ISBN: 978-1-138-12125-6 (hbk)
ISBN: 978-0-367-02978-4 (pbk)

Typeset in Goudy
by Taylor & Francis Books

This book is an outcome of Projet CIRCULEX, Agence Nationale de la Recherche ANR-12-GLOB-0001.

Contents

Illustrations

Figures

Tables

Boxes

Contributors

Maud Borie is an environmental social scientist researching the politics of environmental knowledge and expertise. Using Science & Technology Studies concepts and methods, she recently completed her PhD on IPBES at the University of East Anglia. She has been a visiting fellow with the Harvard STS Program (Fall 2013).

Claudio Chiarolla is Legal Officer in the Traditional Knowledge Division of the World Intellectual Property Organization (WIPO). Prior to joining WIPO, he was Research Fellow at the Institute for Sustainable Development and International Relations (IDDRI) and he has taught postgraduate courses on "Biodiversity Law and Policy" and "Public International Law" at the Paris School of International Affairs (PSIA), SciencePo., Paris. He was also Senior Advisor to the Italian Ministry for the Environment, Land and Sea and acted as a biodiversity negotiator on behalf of the 2014 Italian Presidency of the Council of the European Union.

Daniel Compagnon teaches International Relations at Sciences Po, University of Bordeaux, France. He has contributed to research projects on global governance of climate change, biodiversity negotiations, sustainable development public–private partnerships, and environmental regime complexes. He has published several journal articles and book chapters on these issues and on African politics (https://durkheim.u-bordeaux.fr/Notre-equipe/Chercheurs-et-enseignants-chercheurs/CV/Daniel-Compagnon).

Wolfgang Cramer, environmental geographer and global ecologist, has been Scientific Director at the Mediterranean Institute for Biodiversity and Ecology (IMBE), in Aix-en-Provence, Marseille and Avignon (France), since the establishment of the institute in 2012. The scientific contributions by Cramer's research group (130+ papers) were initially in the area of modelling forest dynamics under climate change. He then began to seek a broader understanding of biosphere dynamics at the global and continental scale, including aspects of natural and human disturbance as well as biodiversity.

Fanny Duperray graduated from the Political Sciences Institute in Lille, and holds a master degree in social sciences from AgroParisTech (Paris institute of technology for life food and environmental sciences). She worked on this publication for her Master thesis, while she was an intern at the ISCC (the CNRS Institute for Communication Sciences) and the CIRAD (CIRAD – Agricultural Research for Development).

Guillaume Futhazar is a PhD student affiliated to the Centre for International and European Research and Study (CERIC) and the "Objectif Terre Méditerranée – Laboratory of Excellence" (OT-Med Labex) based in France. Specializing in international public law and international environmental law, the focus of his current work is on Science Policy Interfaces, fragmentation and regime interaction in the Mediterranean Region.

Marie Hrabanski is a researcher in political sociology from the Centre International de Recherche Agronomique pour le Développement (CIRAD), UMR ART-Dev (France), and she specializes in environmental and agricultural policies at the international level and in developed and developing countries. She has published a large number of articles in this field, focusing particularly on biodiversity and ecosystem service approaches.

Philippe Le Prestre is Professor of Political Science at Laval University (Quebec, Canada). His current research interests revolve around complex systems and the governance of global environmental issues.

Sélim Louafi is a Senior Research Fellow at the Centre International de Recherche Agronomique pour le Développement (Cirad, Montpellier, France). He is interested in science and policy interactions at the global level, in particular in the field of agricultural biodiversity. He is currently a Marie Curie Outgoing Fellow and has been appointed for five years (2014–2019) as a member of the Capacity Building Task Force of the International Platform on Biodiversity and Ecosystem Services (IPBES). He is also member of the Comité Economique, Ethique et Social of the Haut Conseil des Biotechnologies (France)

Sandrine Maljean-Dubois is a researcher at the CNRS and teaches international and European environmental law at the Faculty of Law and Political Sciences of Aix-Marseille University (France). She has edited several books and a large number of articles in this field, focusing in particular on biodiversity and biosecurity and climate change negotiations.

Mohamed Oubenal has a PhD in sociology from Paris-Dauphine University. His research interests include economic sociology, sociology of experts and network analysis. During his postdoctoral fellowship at CIRAD he studied the relationships between experts as well as stakeholders of the Intergovernmental science-policy Platform on Biodiversity and Ecosystem Services (IPBES). As a postdoc in the ISCC, he performed mapping of scientific controversies. He is currently research fellow in sociology at IRCAM in Morocco and associate researcher at the Observatoire des Réseaux Intra et Inter-organisationels (ORIO) in France.

Denis Pesche is a sociologist from the Centre International de Recherche Agronomique pour le Développement (CIRAD), UMR ART-Dev (France) and he specializes in analyzing rural interests groups in the policy making, environmental and agricultural policies, at the international level and in developed and developing countries. He coordinated a research program on ecosystem services from 2009 to 2013. He has published a large number of articles in those fields, focusing particular on biodiversity and ecosystem service approaches.

Annalisa Savaresi is Research and Teaching Fellow in Global Environmental Law at Edinburgh Law School, University of Edinburgh, United Kingdom. She specializes in European, international and comparative environmental law. Her research interests include climate change, forestry, environmental liability, and the interplay between human rights and environmental law. She is a member of the IUCN Commission on Environmental Law and writes for the *Earth Negotiations Bulletin*, reporting on multilateral environmental negotiations on climate change.

Alice B.M. Vadrot is Visiting Research Fellow at the Centre for Science and Policy of the University of Cambridge (UK) and Erwin Schrödinger Fellow of the FWF. In 2014 she published the book *The Politics of Knowledge and Global Biodiversity*. Her current research focuses on the epistemic and political dimensions of institutional arrangement interfacing science and politics.

1 Introduction

Analyzing IPBES functioning within the biodiversity regime complex and beyond

Marie Hrabanski and Denis Pesche

The creation of the Intergovernmental Platform on Biodiversity and Ecosystem Services (IPBES) in 2012 can be seen as key step in international biodiversity governance. Since the 1980s, the neologism "biodiversity" has become a concept that increasingly encompasses various scales of life, from genes to ecosystems. Initially invented by conservation biologists to increase their say in policy decision making (Takacs, 1996), biodiversity is now a key concept for various disciplines, a central issue for some public policies and a specific field in global environmental governance. Biodiversity loss is still, nevertheless, a global trend mainly due to anthropogenic pressures (GBO3, 2010). More recently, the ecosystem service concept has gained influence in framing the issue regarding interactions between ecosystems and human wellbeing. The primary objective of IPBES is, "to strengthen the science-policy interface for biodiversity and ecosystem services for the conservation and sustainable use of biodiversity, long-term human wellbeing and sustainable development" (UNEP, 2012b). More than 20 years after the emergence of the Convention on Biological Diversity (CBD), IPBES now illustrates the slow process of institutionalization of the biodiversity agenda in international governance. The complex institutional landscape of international biodiversity politics within which IPBES is now operating is often seen as deficient, fragmented and unstructured (Le Prestre and Martimort-Asso, 2004; Le Prestre, 2002; Young, 2008). The slow and difficult emergence of IPBES between 2008 and 2012 is both strange and surprising in the highly fragmented biodiversity governance context.

The literature on IPBES has been burgeoning since 2010. Most related publications were authored by scientists eager to promote this option for a new Platform (Larigauderie and Mooney, 2010; Duraiappah and Rogers, 2011; Perrings et al., 2011). Some of this literature on IPBES argues in favor of its role in coherent production and coordination based on scientific findings. Some authors claim to generate policy relevant information based on sound science and targeted observations (Larigauderie and Mooney, 2010). Biodiversity loss is often interpreted as a consequence of the provision of deficient information to decision makers. The science-policy process is often conceptualized as the integration of different elements such as research, monitoring, assessment and policy development with the aim of providing decision makers with suitable information to ensure the best responses (Perrings et al., 2011). While some authors emphasize the co-evolving nature of relations between science and policy (van den Hove, 2007; Jasanoff, 2004; Koetz, Farrell and Bridgewater, 2012), most IPBES-oriented literature shares a vision of more linear interactions between science and policy where scientists and other stakeholders provide knowledge for decision makers.

Another often discussed issue in the literature concerns the diversification of "biodiversity knowledge-holders". Some scholars highlight this issue as a challenge and an opportunity for improving biodiversity governance within IPBES. The key message is often to foster mechanisms able to combine this knowledge diversity within IPBES. But most of those publications were written before the official launch of the Platform in 2013 and combined both analytical and normative assumptions about how IPBES should be or should act (Van den Hove and Chabason, 2009; Turnhout et al., 2012; Goerg, Nesshöver and Paulsch, 2010; Vohland et al., 2011).

More distanced analysis focuses on the complexity and conflicting context of global governance on biodiversity issues (Morin et al., 2015; Vadrot, 2014). Other research has begun focusing on the actual functioning of IPBES concerning the issue of dealing with a plurality of knowledge on biodiversity issues (Soberón and Peterson, 2015; Borie and Hulme, 2015). This book shares this analytical position and is based on the first years of the existence of IPBES. It is structured around two main research questions: How can the emergence of IPBES be explained in a context of fragmentation and complexity of global biodiversity governance? How does IPBES manage tensions between the openness ambition, the knowledge selection process and the production of global assessments by experts? Each chapter provides empirical and analytical elements to address these two main questions.

The next section provides a short description of IPBES and its roots. It explores the literature on global environmental governance and science-policy interactions so as to identify key issues to address the two questions raised in the book.

IPBES – roots and challenges of a new body in global environmental governance

IPBES – a brief history and functions

The challenge of assessing knowledge for policy-making decisions is a large part of any attempt to reduce the numerous uncertainties in global environmental governance. Global environmental assessments have multiplied at the international level since the 1980s. Climate change has been a pioneering element of this trend since the 1988 implementation of a permanent assessment process, i.e. IPCC,[1] several years before the adoption of a an intergovernmental convention, the United Nations Framework Convention on Climate Change

(UNFCCC) (Rio, 1992). Contrary to the work on climate change, no permanent process for the assessment of knowledge on biodiversity questions was available.

The Global Biodiversity Assessment (GBA) (1993–1995) was conducted soon after the signing of CBD. However, in line with its qualification as an independent scientific exercise without any particular political mandate, GBA was not included in the intergovernmental process that started with CBD. The failure of GBA to gain recognition contributed to the motives underlying the launch of a second global biodiversity assessment initiative – the Millennium Ecosystem Assessment (MA) (2001–2005) – but this time care was taken to associate several international organizations in its design and implementation (Pesche et al., 2013), while ensuring that it would remain an "independent" exercise. From 2005, MA and its follow-up were accompanied by a series of activities and initiatives building on the MA findings, such as The Economics of Ecosystems and Biodiversity study (TEEB), the Ecosystem Management Programme (EMP), led by the United Nations Environment Programme (UNEP), and the Programme on Ecosystem Change and Society coordinated by the International Council of Scientific Unions (ICSU) (UNEP, 2010).

MA and its results have influenced research agendas (Carpenter et al., 2009) and prompted the expectation "that there would then be an exploration of the possibility of creating an ongoing IPCC-like process, which should take into account policy makers and stakeholders". MA thus represents an essential milestone in the genesis of IPBES. Meanwhile, in 2005, the French government launched a three-year consultation process in the framework of the International Mechanism of Scientific Expertise on Biodiversity (IMoSEB), and the most important IMoSEB deliverable was the recommendation that a new "intergovernmental [body] with scientific credibility, political legitimacy and relevance" was needed (Babin et al., 2008).

In this context, the debate within CBD on establishing a framework for ensuring the regular production of global assessments of the state of biodiversity worldwide remained focused on how to do so and under what conditions. COP Decision IX/15[2] (2008) was the beginning of a five-year negotiation processes under the auspices of UNEP – involving stakeholders from more than 100 countries, scientific communities and NGOs – with the aim of deciding whether and potentially how to strengthen the science-policy interface for biodiversity. Negotiations leading to the decision that a new body should be established included three Intergovernmental and Multi-Stakeholder Meetings, which took place in 2008, 2009 and 2010, and two Plenary sessions in 2011 and 2012. It officially led to the creation of IPBES in 2012: "The Platform's objective is to strengthen the science-policy interface for biodiversity and ecosystem services for the conservation and sustainable use of biodiversity, long-term human wellbeing and sustainable development" (UNEP, 2012a), "Focusing on government needs and based on priorities established by the Plenary, the Platform responds to requests from governments, including those conveyed to it by multilateral environmental agreements related to biodiversity and ecosystem services as determined by their respective governing bodies".

Four main functions were therefore identified for IPBES. First, "the Platform identifies and prioritizes key scientific information needed for policy makers at appropriate scales and catalyzes efforts to generate new knowledge by engaging in dialogue with key scientific organizations, policymakers and funding organizations, but should not directly undertake new research". Second, "the Platform performs regular and timely assessments of knowledge on biodiversity and ecosystem services and their interlinkages, which should include comprehensive global, regional and, as necessary, subregional assessments and thematic issues at appropriate scales and new topics identified by science and as decided upon by the Plenary". Third, "the Platform supports policy formulation and implementation by identifying policy relevant tools and methodologies (…) for decision makers". Fourth, "the Platform prioritizes key capacity-building needs to improve the science-policy interface at appropriate levels and then provides and calls for financial and other support for the highest-priority needs related directly to its activities" (UNEP, 2012a).

Discussions during the first years of the slow IPBES emergence process were centered on the question of whether and how the science-policy interface should be strengthened. In a complex and sometimes hostile context, IPBES advocates put forward convincing arguments about the relevance of the new Platform. Then, in the two Plenary meetings (2011–2012), the issue of the modalities and institutionalization of the future intergovernmental platform was addressed, highlighting the plurality of visions and interests on the way of governing IPBES. The openness ambition of the Platform was enacted and concerned both the openness to stakeholders (indigenous people, NGOs, private sector, etc.) and to social sciences (economics, anthropology, etc.). The first question addressed in this book concerns the difficulties and controversial context of the IPBES emergence: How can the emergence of IPBES be explained in a context of fragmentation and complexity of global biodiversity governance? The second question explores the impact of the openness strategy on the functioning of the Platform: How does IPBES manage tensions between the openness ambition, the knowledge selection process and the production of global assessments by experts?

Two main scientific-inspired approaches are used to answer these questions. The first approach is derived from the international relations field and deals with IPBES interactions with other Conventions and multilateral environmental agreements (MEAs). The IPBES objective to produce assessments on the biodiversity issue is in a context of growing institutional complexity at the international level. The second approach mobilizes both science and political science studies to explore interactions between science and policy in global environmental governance. Part of this literature focuses on the procedural aspects rather than the substance of the results of so-called science-policy interfaces and suggests that attention should be paid to the social conditions necessary for science to become relevant for policy making. The following section provides further detail on those two approaches.

Beyond IPBES – a new platform in a context of fragmentation and complexity of global biodiversity governance

The idea of IPBES was discussed and the Platform was finally set up in a specific international environmental political setting where there is growing concern about the fragmentation and interplay between environmental regimes (Young, 1996; Young, 2008; Biermann et al., 2009). The rationale for creating IPBES was partly founded on the necessity for coordination and coherence outlined in a complex array of conventions, agreements and institutions dealing with biodiversity issues. The coordination of science has been put forward as a key condition for an "efficient" IPBES (Larigauderie and Mooney, 2010). More broadly, the idea of an IPBES was strongly motivated by the ambition for coordination of the Platform for various purposes: capacity building, scientific assessment, policy recommendations, etc. (UNEP, 2009; Vadrot, 2014). We formulate the hypothesis that this need for coordination, backed by frequent references to the Intergovernmental Panel on Climate Change (IPCC), is a key argument to explain the emergence of IPBES in a potentially hostile context.

Even with IPBES fulfilling its coordination role, the various participants (state, international organizations, NGOs, business, etc.) still approach the "biodiversity problem" differently according to their own norms, technologies, and experience. However, the problems are defined jointly and the criteria for interpreting and accepting knowledge are shared. This gives rise to a challenge regarding IPBES policy coherence with other conventions and MEAs. The objective of IPBES to produce assessments on the biodiversity issue is dealt with in a setting of growing institutional complexity at the international level. Understanding this institutional complexity is a key to gaining further insight into the dynamics of the emergence of IPBES and why this Platform was finally possible.

Since its signing in 1992, the Convention on Biological Diversity (CBD) has sought to frame the different biodiversity issues at the international level. However, this Convention has failed to play this pivotal role between all conventions within the biodiversity regime complex.[3] The creation of IPBES was founded on the question of coordination within this set of international agreements and institutions concerned about biodiversity issues.

CBD also has a Subsidiary Body on Scientific, Technical and Technological Advice (SBSTTA), which is also the case with regard to most other conventions, such as the Ramsar Convention, CITES, CMS, etc., as well as regional or local conventions (Bern Convention). Discussions at the time of the emergence of IPBES sought to build complementarities between these bodies devoted to interactions between science and policy and the overall view was that IPBES would be a competing structure. The topics developed in IPBES may also be closely linked to climate change, and therefore to IPCC and UNFCCC. Therefore, the question of coordination of the recent Platform with other institutions, and the fact that MA was focused on global governance of biodiversity and the environment, more generally sheds light on how IPBES emerged despite this background of growing institutional complexity.

Moreover, the coordination issue is not really a problem of public policy arising from the implementation of overlapping commitments (Velázquez Gomar, Stringer and Paavola, 2014; Velázquez Gomar, 2013). Instead, from an upstream standpoint, it is a problem of coherence among various commitment definitions. In this way, IPBES, as a science-policy interface, aims to provide a means to organize interactions between actors (governments, stakeholders, scientists, etc.) and institutions (CBD, etc.) involved in biodiversity issues at the international level to produce a common framework based on a range of biodiversity knowledge concerns. International relations first proposed to analyze the "institutions possessing norms, decision rules, and procedures which facilitate a convergence of expectations" regarding the notion of regime (Krasner, 1983). However, since the number of international regimes is constantly increasing and each evolves by expanding, some end up overlapping. This leads to the creation of some constellations of regimes, more or less connected to each other, or so-called "regime complexes" (Orsini, Morin and Young, 2013). The different regimes that compose a regime complex can compete, develop synergies or conflict (Young, 1998). Several terms are used in the literature to qualify the links between the different elements of a regime complex. Most of the debate on cooperation between these complexes involves a normative vision geared towards overcoming the "fragmentation problem" to promote coordination and effectiveness of the regime complex. According to Morin and Orsini (2013), "Regime complexity presents governments with an obvious problem of policy coherence" (Morin and Orsini, 2013). Similarly, Alter and Meunier highlight, "the negative consequences of a fragmented regime complex, namely forum-shopping, regime-shifting, strategic inconsistency or inconsistencies in problem-definitions and resulting policies" (Alter and Meunier, 2009). For Keohane and Victor, the regime complex offers advantages such as flexibility and adaptability (Keohane and Victor, 2011). It is probably too early to get a clear answer to those interrogations or about the specific role of IPBES within the biodiversity regime complex. However, the stated ambition of IPBES coordination involves formal and informal mechanisms and processes aimed at achieving policy coherence within the biodiversity regime complex and beyond. One of those processes is the institutional capacity to draw lessons from past experience to design new global environmental governance institutions. In a growing interconnected world, the circulation of norms, ideas and actors has led to policy transfers between countries but also between international organizations, conventions or other elements of the global governance complex (Dolowitz, 2003; Dolowitz and Marsh, 1996; Dolowitz and Marsh, 2000). As shown by neo-institutionalists, policy transfer of a model includes transfer from dena-turation and adaptation, which is the main source of institutional innovation (DiMaggio and Powell, 1991). Similarly, policy transfers are multidimensional and some dimensions may result in a greater degree of convergence, while others do not converge (or diverge), which does not establish a systematic link between coordination, transfer and convergence (Benson and Jordan, 2011; Dumoulin and Saurugger, 2010; Knill, 2005).

According to IPBES and its promoters, the coordination thrust involves the construction of a unified framework, which gathers different knowledge for decision making and is mainly but not solely based on the scientific legitimacy of the experts involved and knowledge represented.

Science-policy interactions within IPBES

One of the four and probably main objectives of IPBES is to produce assessments on biodiversity and ecosystem service issues. As the complete name of IPBES suggests, the Platform aims to manage interactions between "science and policy for people and nature". One common sense usually formulated, and often claimed by scientists, is the view of science revealing the truth for policy makers. There is now a large amount of literature on the complexity of science-policy interactions, particularly political science and science studies (Campbell Keller, 2009).

In mainstream liberal regime theory, "science is understood as a resource that nation-states can use in their negotiations concerning international agreements and has no independent role relative to state interests. Indeed, knowledge is but one of many resources that a state can use when bargaining over international cooperation" (Lidskog and Sundqvist, 2015). A more pluralist and cognitive approach was developed by Peter Haas, stressing the role of policy relevant and consensual knowledge in regime making (Haas, 1990). Haas defines the notion of epistemic community as, "a transnational network of professionals with recognized expertise and competence in a particular domain and an authoritative claim to policy-relevant knowledge within that domain or policy area" (Haas, 1992). This definition highlights the nested nature of links between scientists, policy makers and other stake-holders in the production and diffusion of ideas within the policy process. Nevertheless, some international relations specialists still believe in the independence of science and policy processes (Lidskog and Sundqvist, 2015).

To enhance the overall understanding on the interlinked nature of science-policy interactions, the critical sociology of knowledge (or science and technology studies – STS), as promoted by authors such as Bruno Latour and Sheila Jasanoff, is essential. This branch of literature focuses on the social and economic conditions in knowledge production (Jasanoff, 1987; Jasanoff et al., 1995). Scientific "truths" are relative to a social-historical context and knowledge is always contested. There are competing scientific discourses reflecting the influential interest groups in the political debate – the so-called scientific consensus is primarily a political trade-off. Institutional loyalties, funding ties, personal ambitions and prejudices could explain some of the "scientific" controversies. Scientists are emerged in the society and share its beliefs and feuds, while other social actors, including governmental agencies, interfere with the scientific process. Besides, as scientists address the social demand, particularly from the policy makers, policy-relevant science is always a co-production (Jasanoff, 2004), or what Latour calls "hybridization", to reflect the blurred boundaries between science and policy. Jasanoff argues that there is a need to focus on the social conditions necessary for science to become relevant for policy making (Jasanoff, 2005).

Science and policy are not completely separate processes. Indeed, science is not a neutral, objective and consensual mode of knowledge production, nor are politics and policies partial, subjective and conflicting processes. There is now a relative consensus on the idea that the policy process mobilizes science as a knowledge-based resource and that science is also a social process embedded in various political contexts (Lidskog and Sundqvist, 2015).

Science-policy interactions are crucial for environmental policy, especially at the global level (Dauvergne, 2005). Since the 1980s, scientific assessments of global environmental change have been increasingly mobilized for policy and decision making (Mitchell et al., 2006). For Mitchell and colleagues, assessing knowledge for policy-making decisions is a large part of attempts to reduce the numerous uncertainties in global environmental governance. They define global environmental assessments "as formal efforts to assemble selected knowledge with a view toward making it publicly available in a form intended to be useful for decision making" (Mitchell et al., 2006). These authors question whether global environmental assessments produce effects on decision making and the conditions under which that might occur. Their main conclusion is that three intertwined dimensions underlie the influence of environmental assessment, i.e. credibility, salience and legitimacy. Credibility involves the scientific adequacy of technical evidence and arguments. Salience deals with the relevance of the assessment to the needs of decision makers. Legitimacy reflects the view that information and technology production is respectful of stakeholders' divergent values and beliefs, unbiased in its conduct, and fair in its treatment of views and interest (Mitchell, Clark and Cash, 2006). Since the 2000s, those three elements have become the motto for knowledge management in environmental policy, with consequences on the organizational design of institutions specialized in knowledge selection and assembly. This highlights the accelerating learning processes between applied research, decision makers and practitioners with regard to environmental issues (The Social Learning Group, 2001). In the case of IPBES, salience and legitimacy criteria have raised questions about the diversity of decision makers and other actors concerned by assessed issues. A clear implication for IPBES is the diversification of participants with an effort towards integrating non-state actors as stakeholders in the process.

In the same vein, the science-policy interface (SPI) concept has recently become a common term in environmental policy, with the aim of qualifying organizations or arenas devoted to knowledge selection and assembly for decision making. Rooted in STS literature and research focused on boundary effects of the complexity of science-policy interactions (Cash et al., 2003; Guston, 2001), the science-policy interface notion embodies the idea that science and policy are not mutually exclusive or hermetic categories but rather intersecting domains of human activities which are in co-evolution (van den Hove, 2007). This SPI literature has some normative dimensions and seeks to identify the challenges and conditions for more collaborative models in science-policy interactions (Koetz, Farrell and Bridgewater, 2012; Beck, 2010). The multiplication and diversity of knowledge and actors in the IPBES process suggest that attention should be focused on the possible consequences of this increased diversity in IPBES functioning and products.

As an intergovernmental institution, IPBES tends to produce compromise and build a common vision and operating principles. IPBES can be viewed as an institutional setting where the making of global knowledge is embedded in political and social ordering (Miller, 2004; Jasanoff, 2010). In the case of IPCC, developing country participation, for instance, has been a major challenge since the outset (Agrawala, 1998). The question of expert selection was then a source of friction, especially between developing and industrialized countries – exactly who is allowed and able to take part in transnational expert networks? (Biermann, 2000). The question of the balance between different types of science (earth science, natural science, social science) has been well studied for the purpose of climate change assessments (Hulme and Mahony, 2010). Past GEAs claim to have integrated social science approaches into the process from the very beginning, with the aim of grasping the institutional dimension of human behavior regarding ecosystems (Duraiappah and Rogers, 2011). In IPBES, the expert selection process is crucial for assessments – some improvements have been made in the search for more balanced participation, but social sciences are still largely underrepresented (Montana and Borie, 2015). Countries and scientific disciplines are not the only criteria for knowledge diversification within IPBES.

The specificity of biodiversity issues provides a partial framework for IPBES. Biodiversity is a worldwide problem. Biodiversity loss is typically local and regional and many of the effects are seen at sub-global scales (Duraiappah and Naeem, 2005; Soberon and Sarukhan, 2010). This scalar dimension, reinforced by the cultural dimension of nature/human visions and existing or potential conflicts regarding the commercial use of biodiversity, pave the way for a multifaceted framing process on biodiversity issues, mobilizing a range of different actors. But scalar issues and concepts like global or local are also dependent on the beliefs, actions and practices of the relevant actors: both global and local knowledge are situated (Jasanoff and Martello, 2004). This diversity of knowledge-holders on biodiversity has broadened the scope of traditional environmental assessments focused on scientific knowledge, while recognizing that scientific and experience-based knowledge can be linked, with the involvement of NGOs, the private sector, indigenous people and local communities (Turnhout et al., 2012).

Similar to IPCC, the IPBES intergovernmental functioning and rules of participation for non-state actors define the institutional context in which knowledge is selected and assembled (Hulme, 2010). Concerning non-scientific knowledge, indigenous and local knowledge (ILK) is an important and sensitive issue within IPBES. Representatives of indigenous people request a specific position within the Platform. Since the Rio Declaration, indigenous people have gained ground with respect to asserting their interest on the international scene, particularly in CBD (Mauro and Hardison, 2000). This marks a first step in the recognition of indigenous knowledge (Coombe, 2001) and rights, particularly with article 8(j). CBD above all represents the reaffirmation of national sovereignty over natural resources (Le Prestre, 1999) but, paradoxically, the Convention also marks the willingness of local communities – indigenous or rural – to obtain better control of their areas. Reinforced by the support of some international NGOs, some states and a few scientific communities, representatives of indigenous peoples have demanded the recognition of indigenous knowledge and the institutionalization of their participation in the CBD negotiation process. The issue of the protection of "local knowledge" has become a cross-cutting theme for militants working with developing countries at the end of the 1990s (Roussel, 2005).

The evolving thinking regarding science-policy interactions and the specificities of biodiversity and ecosystem service issues has prompted a strong demand for the integration and representation of diverse and distributed knowledge within IPBES (Beck et al., 2014), with a challenge to bridge scales and epistemologies (Brosius, 2006). Driven by rules and governing principles derived mainly from its intergovernmental status, IPBES is framed by a consensus ideology strengthened by the weight of Western science within it. The increased diversity of participants generates tensions between the openness ambition of inclusion and the selection process of knowledge and experts to produce global assessments. The fact that the main function of IPBES is to produce assessments has led to a continuous exercise of selecting and assembling knowledge from among its growing audience and constituencies.

Structure of the book and the main contents

The first part of the book deals with the status of IPBES in the biodiversity regime complex and beyond, and with the way the platform tries to create coordination between and within the different elements of regime complexes. The first six chapters analyze the creation and governance of IPBES in highlighting the historical and political opportunities and constraints which led to the creation of IPBES. The increased overlap of institutions, agreements and organizations involved in international biodiversity policy promotes the circulation of actors and the transfer of legal, policy, procedural and institutional norms between and within the environmental regime complex. The strong influence of the IPCC model on IPBES fully illustrates this process. However, the creation of IPBES in this institutional complexity context has forced the new Platform to adapt its procedures and operations to the expectations of the various stakeholders involved in biodiversity and its emergence. Thus, the analysis of the genesis and specificities of IPBES rules and procedures highlights the building of a trade-off between a variety of actors with different interests and practices, and who share different ideas on the production of useful knowledge for decision making.

In chapter 2, Philippe Le Prestre and Daniel Compagnon analyze the building process of the biodiversity regime complex and its different elements. Then they analyze the IPBES mechanisms and coordination challenges and show how the coordination challenge entails specific horizontal (between the different conventions and international MEAs) and vertical actions with more or less integrated national biodiversity policies.

The highly politicized character of international biodiversity politics and policies is dependent on the growing diversity of interest among actors around biodiversity issues such as benefit sharing, etc. This has led to many predictions that the attempt to establish a science-policy interface will fail. Why then did the formal establishment of IPBES ultimately succeed? In chapter 3, Alice Vadrot combines both empirical work and theoretical reflection on policy change as a historical development to address this first question. In addition, in the IPBES process, it is becoming increasingly evident that the (monetary) value of biodiversity and ecosystem services could contribute to enhancing the effectiveness and efficiency of related policies and politics.

Chapter 4 describes the governance and status of IPBES. Drawn up as a synthetic presentation of the rules governing IPBES, the aim of this chapter is to provide a broad overview of the IPBES governance dynamics with regard to two aspects – institutional functioning and its core business, namely the production of assessments on biodiversity issues. The authors, Denis Pesche, Guillaume Futhazar and Sandrine Maljean Dubois, also highlight some controversial aspects among participants during the emergence process, including the question of the legal status of IPBES and the inclusion of non-state actors in the intergovernmental process, etc. The chapter also describes IPBES activities through its work programme and gives some elements on its budgetary dimension that tell us about the real priorities of the Platform.

Then chapter 5 provides relevant elements on IPBES global environmental governance strategies, showing how one of the keys to the recognition of IPBES is that it must be based on IPCC procedures to ensure the credibility of the assessments it wishes to produce. In this way, Guillaume Futhazar illustrates the obvious influence IPCC has had on the establishment of IPBES rules of procedure for functional reasons and then highlights the procedural innovations of the Platform and attempts to explain them in the light of the characteristics of the Platform.

Chapter 6, by Daniel Compagnon and Wolfgang Cramer, dovetails perfectly with the previous chapter as it analyzes the process regarding the circulation of actors and norms between IPCC and IPBES by highlighting the experience of an expert involved in the science-policy interfaces of both of these bodies. As the authors point out, inter-scientific dialogue is crucial and there is a necessity to broaden the scope beyond the core natural sciences. This question was already one of the challenges addressed by IPCC but this is even more the case for IPBES.

After analyzing the position of IPBES in its historical, political and organizational environment, the following chapters analyze the actual IPBES functioning process, which aims to integrate a plurality of actors and knowledge. This second part of the book aims to shed some light on the tension between the knowledge and expert selection process and the ambition of openness and inclusiveness, thus hopefully enhancing readers' understanding of the dynamics involved. The chapters highlight the difficulties and contradictions between the aim to increase the diversity of actors, norms and knowledge and the claimed coherence on ways to select and assemble knowledge for decision making.

Indeed, in chapter 7, the IPBES conceptual framework is presented by Maud Borie and Denis Pesche as the main device for building coherence among the huge diversity of knowledge, interests and visions encompassed within the biodiversity science-policy interface. The nickname "Rosetta Stone" highlights the ambition of the IPBES common framework to be an inclusive and multifaceted tool to generate a framework for biodiversity issues. The question remains open about the real capacity of the framework to integrate different knowledge systems. The IPBES conceptual framework is designed to provide guidance for all IPBES activities.

The recognition of the diversity of actors and knowledge within IPBES is substantiated by a policy of inclusion of stakeholders in the intergovernmental process. However, as shown in Chapter 8, Marie Hrabanski, Mohamed Oubenal and Denis Pesche observe an opposite movement to build a unified group of stakeholders as a credible interlocutor within IPBES while also recognizing the diversity of this group. Consequently, some selection processes are developed to limit the diversity of stakeholders and interests, and ensure the seemingly apparent unity of the IPBES stakeholder group.

As shown by Sélim Louafi in chapter 9 on capacity building, most developed countries acknowledge the need to develop the capacity of developing countries, in particular develop a capacity to predict the consequences of current actions on biodiversity, and effectively assess threats related to biodiversity. From that perspective, capacity building activities could simultaneously boost the knowledge capacities of both the providers and recipients, targeted in science-policy activities, while considering biodiversity erosion as mainly the result of conflicting views on the biodiversity issue and how it should be preserved.

Chapter 10 by Claudio Chiarolla and Annalisa Savaresi provides an overview of the most important international legal instruments tailored for the recognition and protection of indigenous and local knowledge systems in global environmental governance, including the UN Convention on Biological Diversity, the UN Framework Convention on Climate Change and relevant human rights instruments. It considers key challenges involved in embracing indigenous and local knowledge systems under IPBES, alongside other knowledge systems, from the institutional, legal, scientific and practical standpoints.

The process used to select the scientific experts is analyzed by Fanny Duperray, Marie Hrabanski and Mohamed Oubenal in chapter 11, and presents features of the first IPBES assessment. The authors put forward explanatory elements to clarify why the pollination theme has been chosen while showing that pollination helps to consolidate the recognition of IPBES as a forum for coordination between several regime complexes, and to strengthen its links with CBD. Through a well-structured empirical framework, the authors also question different dimensions of the recruitment process and show how IPBES framed the long-standing socio-technical controversy on the role of pesticides in the decline of pollinators, notably regarding conflicts of interest that this controversy has given expression to within IPBES.

Finally, Philippe Le Prestre concludes the book by highlighting the main results, and then offers some reflections on IPBES challenges as a scientific governance tool for the biodiversity regime complex and finally on the impact of IPBES on policy making at different levels.

Notes

1 The Intergovernmental Panel on Climate Change was created in November 1988 at the demand of G7 (now G20) by two United Nations organizations, i.e. the World Meteorological Organization (WMO) and the United Nations Environment Programme (UNEP). IPCC presented its first assessment report in 1990, followed by subsequent reports in 1995, 2001, 2007, and the fifth report was prepared in four parts that were presented between September 2013 and November 2014. See Agrawala (1998b).
2 It "[t]akes note of the outcomes of the consultative process towards an international mechanism of scientific expertise on biodiversity" (IMoSEB) (UNEP/CBD/COP/9/INF/34) and

> *welcomes* the agreement of the Executive Director of the United Nations Environment Programme to convene an ad hoc open-ended intergovernmental multi-stakeholder meeting to consider establishing an efficient international science-policy interface on biodiversity, ecosystem services and human well-being, [...] and invites Parties to ensure that appropriate science and policy experts are made available to attend, and also encourages the participation of experts from various regions and disciplines.
>
> (COP Decision IX/15)

3 Usually, what is sometimes referred to as the "biodiversity cluster" pools six multilateral environment agreements, including the Convention on the Conservation of Migratory Species of Wild Animals (CMS), the Convention on International

Trade in Endangered Species of Wild Fauna and Flora (CITES), the Convention on Wetlands of International Importance (Ramsar Convention), the Convention concerning the protection of World Cultural and Natural Heritage (WHC), the International Treaty on Plant Genetic Resources for Food and Agriculture (ITPGRFA) and the Convention on Biological Diversity (CBD) (see Velázquez Gomar, Stringer and Paavola (2014)).

References

Agrawala, S. 1998a. Structural and Process History of the Intergovernmental Panel on Climate Change. *Climatic Change*, 39(4), 621–642.

Agrawala, S. 1998b. Context and Early Origins of the Intergovernmental Panel on Climate Change. *Climatic Change*, 39(4), 605–620.

Alter, K. J. & Meunier, S. 2009. The Politics of International Regime Complexity. *Perspectives on Politics*, 7, 13–24.

Babin, D., Thibon, M., Larigauderie, A., Guinard, S., Monfreda, C. & Brels, S. 2008. Consultative Process towards an IMoSEB–Strengthening the Science-Policy Interface on Biodiversity. *Paris: Centre de coopération Internationale en Recherche Agronomique pour le Développement (CIRAD)*.

Beck, S. 2010. Moving Beyond the Linear Model of Expertise? IPCC and the Test of Adaptation. *Regional Environmental Change*, 11, 297–306.

Beck, S., Borie, M., Chilvers, J., Esguerra, A., Heubach, K., Hulme, M., Lidskog, R., Lövbrand, E., Marquard, E., Miller, C., Nadim, T., Neßhöver, C., Settele, J., Turnhout, E., Vasileiadou, E. & Görg, C. 2014. Towards a Reflexive Turn in the Governance of Global Environmental Expertise. The Cases of the IPCC and the IPBES. *Gaia*, 23, 80–87.

Benson, D. & Jordan, A. 2011. What Have We Learned from Policy Transfer Research? Dolowitz and Marsh Revisited. *Political Studies Review*, 9, 366–378.

Biermann, F. 2000. Science as Power in International Environmental Negotiations: Global Environmental Assessments Between North and South. *Cambridge, MA, Discussion Paper*. 2000–2017: Belfer Center for Science and International Affairs (BCSIA), Environment and Natural Resources Program, Kennedy School of Government, Harvard University.

Biermann, F., Pattberg, P., Van Hasselt, H. & Zelli, F. 2009. The Fragmentation of Global Governance Architectures: A Framework for Analysis. *Global Environmental Politics*, 9, 14–40.

Borie, M. & Hulme, M. 2015. Framing Global Biodiversity: IPBES between Mother Earth and Ecosystem Services. *Environmental Science and Policy*, 54, 487–496.

Brosius, P. 2006. What Counts as Local Knowledge in Global Environmental Assessments and Conventions? In: Reid, W. V., Berkes, F., Wilbanks, T. & Capistrano, D. (eds.) *Bridging Scales and Knowledge Systems: Concepts and Applications in Ecosystem Assessment/Millennium Ecosystem Assessment*. Washington: Island Press.

Campbell Keller, A. 2009. *Science in Environmental Policy*. Cambridge, MA: MIT Press.

Carpenter, S. R., Mooney, H. A., Agard, J., Capistrano, D., Defries, R. S., Dãaz, S., Dietz, T., Duraiappah, A. K., Oteng-Yeboah, A., Pereira, H. M., Perrings, C., Reid, W. V., Sarukhan, J., Scholes, R. J. & Whyte, A. 2009. Science for Managing Ecosystem Services: Beyond the Millennium Ecosystem Assessment. *Proceedings of the National Academy of Sciences of the United States of America*, 106(5), 1305–1312.

Cash, D. W., Clark, W. C., Alcock, F., Dickson, N. M., Eckley, N., Guston, D. H., Jager, J. & Mitchell, R. B. 2003. Knowledge Systems for Sustainable Development. *Proceedings of the National Academy of Science USA*, 100, 8086–8091.

Coombe, R. J. 2001. Recognition of Indigenous Peoples' and Community Traditional Knowledge in International Law. *Thomas Law Review*, 14, 275.

Dauvergne, P. (ed.) 2005. *Handbook of Global Environmental Politics*. Cheltenham: Edward Elgar.

Dimaggio, P. J. & Powell, W. W. 1991. *The New Institutionalism in Organizational Analysis*. Chicago: University of Chicago Press.

Dolowitz, D. & Marsh, D. 1996. Who Learns from Whom: A Review of the Policy Transfer Literature. *Political Studies*, 44(3).

Dolowitz, D. & Marsh, D. 2000. Learning from Abroad: the Role of Policy Transfer in Contemporary Policy Making. *Governance*, 13, 5–24.

Dolowitz, D. P. 2003. A Policy-maker's Guide to Policy Transfer. *The Political Quarterly*, 74, 101–108.

Dumoulin, L. & Saurugger, S. 2010. Les policy transfer studies: analyse critique et perspectives. *Critique Internationale*, 3, 9–24.

Duraiappah, A. K. & Naeem, S. 2005. *Millennium Ecosystem Assessment: Ecosystems and Human Well-being: Biodiversity Synthesis*. Washington, DC: World Resource Institute.

Duraiappah, A. K. & Rogers, D. 2011. The Intergovernmental Platform on Biodiversity and Ecosystem Services: Opportunities for the Social Sciences. *Innovation: The European Journal of Social Science Research*, 24, 217–224.

GBO3 2010. *Global Biodiversity Outlook 3*. Montreal: Secretariat of the Convention on Biological Diversity.

Görg, C., Neßhöver, C. & Paulsch, A. 2010. A New Link between Biodiversity Science and Policy. *Gaia*, 19(3), 183–186.

Gomar, J. O. V., Stringer, L. C. & Paavola, J. 2014. Regime Complexes and National Policy Coherence: Experiences in the Biodiversity Cluster. *Global Governance*, 20, 119–145.

Guston, D. H. 2001. Boundary Organizations in Environmental Policy and Science: An Introduction. *Science, Technology & Human Values*, 26, 399–408.

Haas, P. M. 1990. *Saving the Mediterranean: The Politics of International Environmental Cooperation*. New York: Columbia University Press.

Haas, P. M. 1992. Introduction: Epistemic Communities and International Policy Coordination. *International Organization*, 46, 1–35.

Hulme, M. 2010. Problems with Making and Governing Global Kinds of Knowledge. *Global Environmental Change*, 20, 558–564.

Hulme, M. & Mahony, M. 2010. Climate Change: What do we Know about the IPCC? *Progress in Physical Geography*, 34, 705–718.

Jasanoff, S. 2004. *States of Knowledge: The Co-production of Science and Social Order*. London, New York: Routledge.

Jasanoff, S. 2005. *Designs on Nature: Science and Democracy in Europe and the United States*. Princeton, NJ: Princeton University Press.

Jasanoff, S. 2010. A New Climate for Society. *Theory, Culture & Society*, 27, 233–253.

Jasanoff, S., Markle, G., Petersen, J. C. & Pinc, T. (eds.) 1995. *Handbook of Science and Technology Studies*. Thousand Oaks, London, New Delhi: Sage.

Jasanoff, S. & Martello, M. L. 2004. *Earthly Politics: Local and Global in Environmental Governance*. Cambridge, MA: MIT Press.

Jasanoff, S. S. 1987. Contested Boundaries in Policy-Relevant Science. *Social Studies of Science*, 17, 195–230.

Keohane, R. O. & Victor, D. G. 2011. The Regime Complex for Climate Change. *Perspectives on Politics*, 9, 7–23.

Knill, C. 2005. Introduction: Cross-national Policy Convergence: Concepts, Approaches and Explanatory Factors. *Journal of European Public Policy*, 12, 764–774.

Koetz, T., Farrell, K. & Bridgewater, P. 2012. Building Better Science-policy Inter-faces for International Environmental Governance: Assessing Potential within the Intergovernmental Platform for Biodiversity and Ecosystem Services. *International Environmental Agreements: Politics, Law and Economics*, 12, 1–21.

Krasner, S. D. 1983. *International Regimes*. Ithaca, NY: Cornell University Press.

Larigauderie, A. & Mooney, H. A. 2010. The Intergovernmental Science-policy Platform on Biodiversity and Ecosystem Services: Moving a Step Closer to an IPCC-like Mechanism for Biodiversity. *Current Opinion in Environmental Sustainability*, 2, 9–14.

Le Prestre, P. 1999. La Convention sur la diversité biologique: vers un nouvel ordre biologique international? *Nature Sciences Sociétés*, 7, 64–71.

Le Prestre, P. & Martimort-Asso, B. 2004. Les questions soulevées par le système de gouvernance internationale de l'environnement. *Rapport IDDRI*.

Le Prestre, P. G. 2002. *Governing Global Biodiversity: The Evolution and Implementation of the Convention on Biological Diversity*. Aldershot, England: Ashgate.

Lidskog, R. & Sundqvist, G. 2015. When Does Science Matter? International Relations Meets Science and Technology Studies. *Global Environmental Politics*, 15, 1–20.

Mauro, F. & Hardison, P. D. 2000. Traditional Knowledge of Indigenous and Local Communities: International Debate and Policy Initiatives. *Ecological Applications*, 10, 1263–1269.

Miller, C. 2004. Climate Science and the Making of a Global Political Order. *In*: Jasanoff, S. (ed.) *States of Knowledge: The Co-production of Science and the Social Order*. London: Routledge

Mitchell, R. B., Clark, W. C. & Cash, D. W. 2006a. Information and Influence. *In*: Mitchell, R. B., Clark, W. C., Cash, D. W. & Dickson, N. M. (eds.) *Global Environmental Assessments: Information and Influence*. Cambridge, MA: MIT.

Mitchell, R. B., Clark, W. C., Cash, D. W. & Dickson, N. M. (eds.) 2006b. *Global Environmental Assessments: Information and Influence*. Cambridge, MA: MIT.

Montana, J. & Borie, M. 2015. IPBES and Biodiversity Expertise: Regional, Gender, and Disciplinary Balance in the Composition of the Interim and 2015 Multidisciplinary Expert Panel. *Conservation Letters*, 1–5.

Morin, J.-F., Louafi, S., Orsini, A. & Oubenal, M. 2015. Boundary Organizations in Regime Complexes. A Social Network Assessment of IPBES. *International Studies Association*. New Orleans.

Morin, J.-F. & Orsini, A. 2013. Regime Complexity and Policy Coherency: Introducing a Co-adjustments Model. *Global Governance: A Review of Multilateralism and International Organizations*, 19, 41–51.

Orsini, A., Morin, J.-F. & Young, O. 2013. Regime Complexes: A Buzz, a Boom, or a Boost for Global Governance? *Global Governance: A Review of Multilateralism and International Organizations*, 19, 27–39.

Perrings, C., Duraiappah, A., Larigauderie, A. & Mooney, H. 2011. The Biodiversity and Ecosystem Services Science-Policy Interface. *Science*, 331, 1139–1140.

Pesche, D., Méral, P., Hrabanski, M. & Bonnin, M. 2013. Ecosystem Services and Ppayments for Environmental Services: Two Sides for the Same Coin? *In:* Muradian, R. & Rival, L. (eds.) *Governing the Provision of Ecosystem Services.* Springer.

Roussel, B. 2005. Savoirs locaux et conservation de la biodiversité: renforcer la représentation des communautés. *Mouvements*, 82–88.

Soberón, J. & Peterson, A. T. 2015. Biodiversity Governance: A Tower of Babel of Scales and Cultures. *PLoS Biol*, 13, e1002108.

Soberon, J. M. & Sarukhan, J. K. 2010. A New Mechanism for Science-policy Transfer and Biodiversity Governance? *Environmental Conservation*, 36, 265–267.

Takacs, D. 1996. *The Idea of Biodiversity: Philosophies of Paradise.* Baltimore, London: The Johns Hopkins University Press.

The Social Learning Group 2001. *Learning to Manage Global Environmental Risks, Volume 1: A Comparative History of Social Responses to Climate Change, Ozone Depletion, and Acid Rain.* Cambridge: MIT Press.

Turnhout, E., Bloomfield, B., Hulme, M., Vogel, J. & Wynne, B. 2012. Listen to the Voice of Experience. *Nature*, 488.

UNEP 2009. Preliminary Gap Analysis for the Purpose of Facilitating the Discussions on How to Strengthen the Science-policy Interface. Governing Council of the UNEP, UNEP/GC.25/INF/30.

UNEP 2010. Busan Outcome. UNEP/IPBES/3/L.2/Rev.1. www.ipbes.net/meetings/ Documents/ipbes3/K1030396-IPBES-3-L.2Rev1.pdf.

UNEP 2012a. Report of the Second Session of the Plenary Meeting to Determine Modalities and Institutional Arrangements for an Intergovernmental Science-Policy Platform on Biodiversity and Ecosystem Services. UNEP/IPBES.MI/2/9. (ed.). 16–21 April 2012, Panama City.

UNEP 2012b. Resolution on the Intergovernmental Science-Policy Platform on Biodiversity and Ecosystem Services (IPBES). United Nations Environment Programme (UNEP).

Vadrot, A. B. M. 2014. *The Politics of Knowledge and Global Biodiversity.* Abingdon, New York: Routledge.

Van den Hove, S. 2007. A Rationale for Science-policy Interfaces. *Futures*, 39, 807–826.

Van den Hove, S. & Chabason, L. 2009. *The Debate on an Intergovernmental Science-Policy Platform on Biodiversity and Ecosystem Services (IPBES): Exploring Gaps and Needs.* Paris: IDDRI, Idées pour un Débat n°9.

Velázquez Gomar, J. O. 2013. *Co-evolution of Regime Complexes and National Policy Coherence: The Case of the Cluster of Biodiversity-related Conventions and National Implementation Systems in Latin America and the Caribbean.* University of Leeds.

Velázquez Gomar, J. O., Stringer, L. C. & Paavola, J. 2014. Regime Complexes and National Policy Coherence: Experiences in the Biodiversity Cluster. *Global Governance*, 20, 119–145.

Vohland, K., Mlambo, M. C., Horta, L. D., Jonsson, B., Paulsch, A. & Martinez, S. I. 2011. How to Ensure a Credible and Efficient IPBES? *Environmental Science & Policy*, 14, 1188–1194.

Young, O. R. 1996. Institutional Linkages in International Society: Polar Perspectives. *Global Governance*, 2, 1–23.

Young, O. R. 1998. *Creating Regimes Arctic Accords and International Governance.* Ithaca, NY: Cornell University Press.

Young, O. R. 2008. The Architecture of Global Environmental Governance: Bringing Science to Bear on Policy. *Global Environmental Politics*, 8, 14–32.

2 IPBES and governance of the international biodiversity regime complex

Philippe Le Prestre and Daniel Compagnon

This chapter outlines the institutional and political context of the Platform's future operations and identifies the nature of the coordination challenge that IPBES faces, as well as the initial measures that it has adopted in order to meet them.

Although many analysts stress the "fragmentation" of international environmental governance, we prefer the notion of "regime complex" in order to highlight interactions among its components, which may generate a chaotic, contradictory global regulatory environment or work in a more collaborative and complementary fashion. After a brief theoretical overview and a subsequent attempt to delineate the biodiversity regime complex (hereafter referred to as "the Complex"), the chapter proceeds with a review of the role IPBES could play in bringing scientific coherence to the Complex, given that the latter already harbors several science policy interfaces.

Of course, this raises the question of the complementary or competitive relationship between these decades-old institutions and the newly created IPBES. As a latecomer established outside the perimeter of existing conventions by the UN General Assembly, the Platform is not necessarily well placed to coordinate existing science/policy interfaces, especially when taking international biodiversity politics into account. However, with its broader mandate that includes the controversial ecosystem service evaluation, IPBES could help integrate various knowledge systems, provide a wider and consolidated perspective, and therefore help tackle issues that have impeded effective policy making in the biodiversity regime complex since the early 1990s.

The first section clarifies the notion of regime complex and applies it to the biodiversity domain, in relation to other possible conceptualizations. It then proceeds by identifying some of the coordination challenges, especially from the science/policy interface angle. The second section will examine more specifically how that challenge has been met so far, and how IPBES can operate and gain legitimacy within the biodiversity regime complex through its interactions with the other components of the complex.

Emergence and structure of a biodiversity regime complex

Having assessed the relevance of the regime complex approach in the first sub-section, we will apply it to biodiversity and try to map out the biodiversity regime complex. A third subsection will focus on science-policy interfaces within such complexes and identify the main challenges facing IPBES in its search for effectiveness and legitimacy in this densely populated field.

The notion of regime complex

Fragmentation has become a feature of global governance as a whole (Biermann, Patterberg & Van Asselt, 2009) and applies also to specific issue areas such as climate change, biodiversity and sustainable development. The notion of fragmentation attempts to capture the diversity of institutional elements and actors such as various intergovernmental regimes, intergovernmental organizations (including treaty secretariats), and transnational governance mechanisms involving non-state actors (Zelli & van Asselt, 2013). It has a normative underpinning, however – particularly in international public law – as it is suspected of impeding effectiveness and coherence, where integration would be preferable. There is an implicit unattainable ideal of a unified international legal order under a UN umbrella.

Attempts to overcome this fragmentation and the potential overlaps and conflicts it generates have given rise to a growing body of research on regime interplay (Young, 2002) or interactions within or beyond the environmental domain. It includes both horizontal interactions at the global level and vertical interactions between a global regime and regional governance systems (Oberthür & Gehring, 2006). Comparative case-study analyses have established that these interactions are often conflict-prone as the norms and institutional mechanisms of these regimes can differ significantly (see case studies in Oberthür & Stokke, 2011). Therefore, the management of these complex interactions would aim at promoting synergy, fostering efficiency and enhancing justice and equity (Oberthür, 2009).

One proposed option to increase synergy is the "clustering" of compatible multilateral environmental agreements (MEAs), i.e. the "combination, grouping, consolidation, integration or merger of MEAs or parts thereof" (Oberthür, 2002: 335). Clustering is meant to counterbalance the excessive fragmentation and favor policy integration, but it remains largely an intellectual project. It has been effectively implemented only for the chemical and hazardous waste-related conventions (Velázquez Gomar, 2014) although a cluster of biodiversity-related conventions is also mentioned. Another way of making sense of the high degree of fragmentation of the international policy framework is to characterize it as a "regime complex", defined as "a network of three or more international regimes that relate to a common subject matter; exhibit overlapping membership; and generate substantive, normative, or operative interactions" (Orsini, Morin & Young, 2013: 29).

It was first used in global environmental governance to characterize the plant genetic resources set of international regulations, with the complex being "an array of partially overlapping and non-hierarchical institutions governing a particular issue-area" (Raustiala & Victor, 2004: 279). Working in exactly the same issue area, Andersen proposes the notion of "regime constellations" that "refers to the entirety of international norms and rules pertaining to an issue area that result from all the international regimes pertaining to that issue area, the overlaps of their functional scopes, and the interaction between them" (Andersen, 2008: 39–40). Although the astronomical image is a powerful metaphor, it does not add analytical clarity to our discussion.

Whereas clustering and interplay management are deliberate attempts to improve coordination among overlapping MEAs, the notion of regime complex offers an analytic lens to help understand the dynamics of a crowded field: "looking through this new conceptual lens, regime complexes may appear where previously analysts saw only individual decomposable regimes" (Raustiala & Victor, 2004: 306). Regime complexes "should be viewed not as ideal constructions but outcomes that emerge from real-world political, organizational, and informational constraints" (Keohane & Victor, 2011: 19).

Keohane and Victor, in analyzing the complex of climate change, adopt a broader understanding of the concept. For Orsini, Morin & Young (2013), a) the constitutive elements are regimes in interaction, whereas these elements are more diverse for Keohane and Victor and are loosely connected; b) the complex is structured by a specific subject matter that the constitutive regimes have in common, whereas Keohane, Victor and Raustalia only mention an issue area in a broad sense (such as climate change). "Regime complexes are marked by connections between the specific and relatively narrow regimes but the absence of an overall architecture or hierarchy that structures the whole set" (Keohane & Victor, 2011: 8). The regime complex is positioned between two extremes on a continuum: an integrated regime imposing regulations with hierarchical rules, and a collection of institutions with no identifiable core and interlinkages or only weak ones.

Kehoane and Victor identify three factors that explain the emergence of regime complexes:

- Diverging interests of key players prevent integrated regulatory frameworks;
- Uncertainty about the distribution of costs favor "clubs" of members with the same interests on specific issues;
- Issue linkage with regimes outside the issue area of the complex favor specific arrangements.

These factors interconnect in the making of the regime complex for climate change with two other factors: the intricate problem structure (the diversity of issues implies creating "an array of different cooperation games") and the political difficulties of implementing policies affecting the individual behavior of a myriad of actors. In this respect, biodiversity is very similar to climate change, and the emergence of a biodiversity complex does not come as a surprise.

Whereas "fragmentation" is usually seen as a predicament, "the emergence of a climate change regime complex, rather than an integrated, comprehensive climate change regime does not necessarily provide reasons to despair" (Keohane & Victor, 2011: 14). For Keohane and Victor, regime complexes offer some flexibility across issues and can adapt to changes of surrounding conditions – in global economy and international politics – more easily than more rigid, fully integrated regimes with hierarchically enforced rules, such as the Kyoto Protocol. It is easier to manage conflicts and synergies, within the complex and at its borders with other areas of international cooperation. Regime complexes allow for some emulation in the search for best practices, distinct from (although also sometimes fostering) the "forum shopping" practiced by governments willing to escape specific regulations. They also facilitate experimentations (e.g. on REDD+), including the use of economic instruments such as transborder adjustment taxes when working in cooperation with other regimes, i.e. the World Trade Organization (WTO) in the latter case. Through these experiments, we see a convergence of policy models at national and subnational levels taking over from conventional international cooperation. Coordination or integration in a regime complex is not entirely left to centralized mechanisms.

Although Keohane and Victor do mention NGOs and firms as sources of regulatory mechanisms, their approach is predominantly state-centered. Abbott broadened the regime complex scope by taking fully non-state actors and transnational governance initiatives into account when mapping the climate change governance complex (Abbott, 2011). Indeed, it makes sense to combine private carbon transparency (e.g. the Carbon Disclosure Project) and certification initiatives with the cap-and-trade intergovernmental regime established under the Kyoto Protocol in the same analytical framework. Abbott's governance complex includes transnational actors engaged in "regulatory standard-setting" and other activities that contribute to the "polycentric governance" of climate change.

Similarly, some forms of polycentric governance of biodiversity encompass resource use certification (e.g. Forest Stewardship Council and Marine Stewardship Council) and public/private partnerships, as well as various private conservation initiatives. However, since IPBES is intergovernmental, its primary impact will be felt by similar international institutions. It would be a different and ambitious project to attempt to measure its impact on knowledge communities, private actors and NGOs. We therefore focus on the regime complex.

Mapping the biodiversity regime complex

There are more than 150 MEAs related to biodiversity, some regional, global and issue-related, while others are general. Some of the most important ones

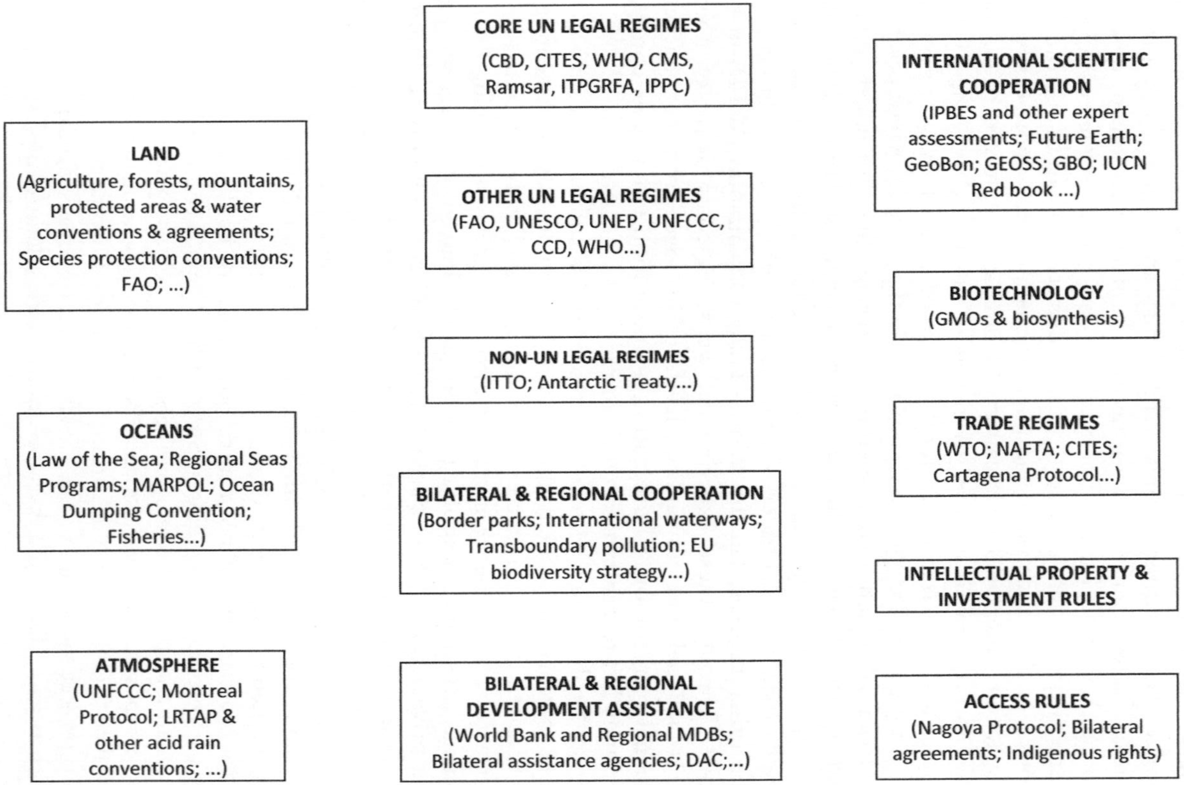

INTERNATIONAL SCIENTIFIC COOPERATION
(IPBES and other expert assessments; Future Earth; GeoBon; GEOSS; GBO; IUCN Red book ...)

BIOTECHNOLOGY
(GMOs & biosynthesis)

TRADE REGIMES
(WTO; NAFTA; CITES; Cartagena Protocol...)

INTELLECTUAL PROPERTY & INVESTMENT RULES

ACCESS RULES
(Nagoya Protocol; Bilateral agreements; Indigenous rights)

CORE UN LEGAL REGIMES
(CBD, CITES, WHO, CMS, Ramsar, ITPGRFA, IPPC)

OTHER UN LEGAL REGIMES
(FAO, UNESCO, UNEP, UNFCCC, CCD, WHO...)

NON-UN LEGAL REGIMES
(ITTO; Antarctic Treaty...)

BILATERAL & REGIONAL COOPERATION
(Border parks; International waterways; Transboundary pollution; EU biodiversity strategy...)

BILATERAL & REGIONAL DEVELOPMENT ASSISTANCE
(World Bank and Regional MDBs; Bilateral assistance agencies; DAC...)

LAND
(Agriculture, forests, mountains, protected areas & water conventions & agreements; Species protection conventions; FAO; ...)

OCEANS
(Law of the Sea; Regional Seas Programs; MARPOL; Ocean Dumping Convention; Fisheries...)

ATMOSPHERE
(UNFCCC; Montreal Protocol; LRTAP & other acid rain conventions; ...)

Figure 2.1 Preliminary mapping of the biodiversity regime complex

from a conservation perspective are the Convention on the Conservation of Migratory Species of Wild Animals (CMS), the Convention on International Trade in Endangered Species of Wild Fauna and Flora (CITES), the Convention on Wetlands of International Importance (Ramsar Convention), the Convention Concerning the Protection of the World Cultural and Natural Heritage (WHC), the International Treaty on Plant Genetic Resources for Food and Agriculture (ITPGRFA) and finally the Convention on Biological Diversity (CBD), i.e. a group of MEAs often referred to as the "biodiversity cluster" (Velázquez Gomar, Stringer & Paavola, 2013: 3).

Inter-treaty coordination was delegated to secretariats and supported by the United Nations Environment Programme (UNEP) and the International Union for the Conservation of Nature (IUCN). The Liaison Group of Biodiversity-related Conventions (BLG) was established in 2004 to enhance coherence and cooperation in the implementation of conventions (CBD Decision VII/26 par. 2).[1] This coordination achieved some results in technical matters, nomenclature harmonization and joint missions (Jardin 2010), however many overlaps remain that have not been addressed. The need for coherence of various policy frameworks has been underlined by the failure to reach the 2010 Biodiversity target adopted at CBD-COP 6 in 2002. Although it was endorsed by the World Summit on Sustainable Development in Johannesburg and later incorporated into the strategic plans of most of the cluster conventions, it had limited impact outside of the CBD. A similar challenge is to cluster coordination by achieving the new Aichi Biodiversity Targets set at CBD COP10 in Nagoya. However, the biodiversity cluster is only one of the possible thematic groupings of MEAs (Le Prestre & Martimort-Asso, 2004: 36), and it is not sure that excluding trade, property rights and development issues from the cluster will really be effective. In spite of its limited human and financial resources, the role of UNEP in coordinating biodiversity policies should not be underestimated (Bauer, 2009), especially when it comes to marshalling scientific expertise (see below).

With the threefold objective of biodiversity conservation, sustainable use and fair and equitable sharing of benefits arising from the utilization of genetic resource, the Convention on Biological Diversity is at the core of the complex. However, the CBD coexists with a number of other international treaties on resource and nature conservation, is not formally a framework convention,[2] and is definitely not the umbrella convention originally proposed by the USA and some NGOs. Consequently it has no institutional precedence over other biodiversity-related MEAs. Yet it has a large scope, certainly broadened with the issues discussed under article 8j[3] and with the ecosystem service approach. Both its Cartagena Protocol on Biosafety, i.e. on transfer of GMOs across national boundaries, and its Nagoya Protocol on Access and Benefit Sharing overlap the TRIPS[4] in WTO (Rosendal, 2006) and other FAO and WIPO property rights regimes.[5] Within CBD itself, Rosendal defines three clusters (biodiversity conservation, forest management, access and benefit sharing) that also involve segments of other regimes. Therefore the biodiversity regime complex is not confined to the CBD, its protocols or even the biodiversity cluster. For example, the World Bank-administered Global Environmental Facility, i.e. the main financial arm for the implementation of the CDB, is not institutionally linked to bodies of the Convention, but it still belongs to the complex.

Indeed, the boundaries of the complex and its internal structure remain a matter of debate. For example, Raustiala and Victor see a more specific regime complex for genetic resources rather than a biodiversity complex as a whole. In this complex, they include five elemental regimes[6] (only four for Andersen, 2008): three regimes specifically for agriculture, one for trade-related property rights under WTO and CBD. In addition, Raustiala and Victor see the US national legislation on intellectual property rights and industrial patents as an integral part of the complex (Raustiala & Victor 2004: 283–284). Therefore, the structure of the regime complex and its boundaries are defined by a set of issues to be regulated (in this case the ownership of and attribution of benefits from plant genetic resources). Since the regime complex is an analytical lens, the boundaries of the regime vary according to the issue addressed. They are not arbitrarily predefined.

In relation to conservation, interactions between the CBD, focused on protected areas and conservation plans, the Ramsar Convention also devoted to protecting habitats, and the CITES regime targeting trade as one factor of biodiversity erosion, but also the quasi-regime on sustainable forestry[7] are obvious picks. However, many international regimes partially address or affect biodiversity conservation and management. Within the climate change regime, the REDD+ negotiations have explicitly included biodiversity pre-servation in Megadiverse countries as a core objective, while Land Use, Land-Use Change and Forestry (LULUCF) activities also have a direct bearing on ecosystem management. Therefore, these two issue areas effectively connect the climate change and biodiversity regime complexes. Other nonbiodiversity conventions with significant but partial overlaps are the United Nations Convention to Combat Desertification (UNCCD) and the UN Convention on the Law of the Sea (UNCLOS). On the frontiers of the biodiversity domain, overlaps and interactions with other areas of international coop-eration (e.g. integration of the new Aichi targets into the recently adopted sustainable development goals (goals 14 and 15) that will shape development policies in the next decade) constantly redefine the boundaries of the biodi-versity regime complex, to which IPBES is a recent addition. The Platform, which is institutionally separate from the CBD, will also impact several other biodiversity-related MEAs.

Of course, the biodiversity complex extends across scales, with its vertical interactions with more or less integrated national biodiversity policies, as much as horizontal interplay between MEAs (Gomar, Stringer & Paavola, 2013; Andersen, 2008). It is important to include non-state actors, given the growing importance of transnational governance initiatives, in particular public/private partnerships (PPP) in sustainable forestry, wildlife conservation and carbon offset and other climate change related issues. Their role is much broader and farther reaching than being mere observers in conferences of the parties and other international negotiations, although their participation in these forums is far from anecdotal (Orsini & Compagnon, 2013; Orsini, Morin & Young, 2013). The same remark applies to environmental NGOs. IUCN was a strong supporter of many MEAs at an early stage, including CBD, and plays a role in monitoring their implementation.[8] Its assembly every four years, the World Conservation Congress, is an important forum of deliberation on biodiversity policies. A number of other global NGOs such as WWF are also strongly involved in the biodiversity field, including in the scientific debate on what type of knowledge really matters.[9]

Science-policy interfaces and the coordination challenge

From the outset, science has been at the core of biodiversity conventions, as the term itself was forged by scientists,[10] and the equally intangible notion of "biodiversity erosion" relies on biologists' assessments. Yet, although there is a global consensus on the acceleration of the biodiversity erosion (the "sixth extinction"), the CBD text is not grounded on robust biological science (it is instead a list of political and economic principles). No global scientific assessment was carried out before negotiations took place in 1991–1992, thus reducing the influence of scientists in defining the aims of the convention.

Unfortunately, the CBD Subsidiary Body on Scientific, Technical and Technological Advice (SBSTTA) has succeeded neither as a catalyst of new research programs nor as a broker between scientists and decision makers. It has apparently been less effective than its namesake in the UNFCCC (see chapter 7), or even its more discreet equivalents in the biodiversity cluster.[11] Most SBSTTA delegates have a governmental official profile[12] and few true biodiversity experts attend its meetings (about 7% of the delegates are from academic and research institutions). Its debates – sometimes with little if any science content – tend to echo or anticipate those of the COP (in the form of drafted decisions), while failing to provide the sound scientific and technical advice specified under its mission. Yet building a scientific consensus is not its role, and COP decisions have significantly restricted its activities, e.g. preventing SBSTTA from launching its own assessments (Le Prestre, 2002: 104). Some careful observers argue that the political nature of SBSTTA is inseparable from it being a science policy interface and that SBSTTA has partially fulfilled its function (Koetz et al., 2008). SBSTTA also encourages the "socialization of delegates of developing countries and industrialized countries alike into the science and norms of the regime" (Le Prestre, 2002: 86).

Part of the difficulty was that the mandate was too ambitious, while policy-relevant scientific and technical knowledge was not always available. The fulfillment of this mandate was also at the center of a political conflict between developed and developing countries. The latter were originally hostile to the creation of SBSTTA (as an institution likely to be dominated by Western scientists), but they were willing to use it to further their own interests. Meanwhile, Western countries were trying to limit its discussions to natural sciences, while preventing topics such as benefit sharing and property rights linked to the biotechnology industries from appearing on the agenda. Besides, social sciences have remained underrepresented among the experts involved in SBSTTA activities, although it is obvious that their involvement is increasingly necessary to cope with implementation problems, especially involving community conservation and local knowledge.

The CBD Secretariat publishes the Global Biodiversity Outlook, which summarizes biodiversity erosion data and trends as compared to the COP objectives. The first report was published in 2001 and the fourth edition of the GBO was released for COP12 in 2014. Although the preparation of this fourth report was reportedly overseen by the SBSTTA Bureau, and a draft of the report was debated at the 18th meeting of SBSTTA (according to the CBD website), the bulk of the work was done by an *ad hoc* scientific advisory group piloted by the Secretariat. The Secretariat also plays an important role in drafting the agenda and the documents submitted to SBSTTA for discussion (Koetz et al., 2008), and overall can be seen as a component of the science-policy interface within the CBD, in potential competition with SBSTTA itself (Le Prestre, 2002: 107).

Historically, global biodiversity assessments were produced outside of the CBD and were sometimes held in suspicion by some parties. UNEP – a strong advocate since its inception of backing up policy with scientific expertise – produces the Global Environmental Outlook, and commissioned the Global Biodiversity Assessment (GBA), which was conducted between 1993 and 1995 with Global Environmental Facility (GEF) financial support. In spite of a strict peer-reviewed process and the good scientific quality of the outcome, the GBA had no noticeable policy impact. UNEP was also tasked with coordinating the Millennium Ecosystem Assessment (MA) between 2001 and 2005, although the scientific work was piloted by an *ad hoc* secretariat backed by an *ad hoc* multi-stakeholder network. The MA attracted a wider audience than the GBA because its structure allowed multi-stakeholder participation, representatives from the biodiversity cluster conventions were more closely involved, and it adopted a different approach centered on ecosystems and ecosystem services. However, beyond promoting the ecosystem service approach, it still had a limited impact on decision making at the regime and national levels. It was not directly commissioned by governments, which were not involved in the proceedings. In particular, the MA did not spur new actions to meet the 2010 biodiversity target. This prompted MA promoters to advocate the creation of an "IPCC-like mechanism",[13] i.e. organized in an intergovernmental manner so that governments would claim ownership of the end product – a move that ultimately led to the creation of IPBES (Larigauderie & Mooney, 2010).

From this brief overview, it appears that the policy relevance of any bio-diversity assessment should not be taken for granted. Scientists need to take into account how policy-making works, including the politics involved, sometimes behind the scene. The politicization of the SBSTTA debates and the absence of institutional links between the previous assessment processes and the CDB jeopardized the transmission pathway between scientific circles and decision makers. IPBES, being a permanent intergovernmental structure with a comprehensive mandate, is certainly better equipped to influence policy making in the future, at global, regional and national levels, especially given its multi-scalar strategy for producing assessments (Beck, 2014: 84). It has the "potential to address the identified institutional mismatches" in the global biodiversity science-policy interface, but it also dispels some features of the outdated "linear model of science-policy interrelations" (Koetz, Farrel & Bridgewater, 2011).

In this respect, mimicking IPCC could prove counterproductive when some voices criticize the policy relevance, or even the legitimacy of the latter's reports (see chapter 7). IPBES, however, integrates several lessons learned from IPCC, notably the need to have governments on board from the beginning, with significant representation of experts from the developing South, the adoption of a multi-stakeholder perspective, the inclusion of capacity-building, and recognition of the cultural context of knowledge. In addition, potential overlaps between the IPBES process and SBSTTA must be carefully managed to prevent political deadlocks. Similarly, it is important that IPBES assessment reports and the Global Biodiversity Outlook are not perceived as redundant or contradictory. Practical coordination could be provided by UNEP, the CBD Secretariat (the most logical solution given its current role), the IPBES Secretariat, or all of them jointly, but turf wars should be avoided at all costs.

IPBES as a potentially integrative mechanism within the biodiversity regime complex

Given the organization is at the early stage of its development, it is premature to draw any firm conclusion regarding the ability of IPBES to fulfill its missions. Rather, we could attempt to identify the challenges it faces, sort out the initiatives that have taken place so far in meeting them, and identify current opportunities and obstacles.

One of these challenges pertains to the coordination and integration of the behavior of the components of the regime complex. Coordination can be understood as ensuring that elements of the complex will not work at cross-purposes. Integration, on the other hand, goes further and points to the capacity to facilitate a common definition of the problem and of its solutions, while helping the Complex improve the science-policy interface. In other words, IPBES is to facilitate an emergent property, namely convergence of the scientific as well as policy-related dimensions of the Complex. While coordination may not lead to convergence, the latter will substantially benefit from the former.

Coordination issues are indeed pivotal to the operation of a regime complex (see above). Although not a research organization per se, IPBES is designed to promote a scientific consensus on the extent of biodiversity loss and on the challenges it presents, while stabilizing biodiversity-related knowledge with the aim of improving the science-policy interface. As such, it responds to one of the problems that Keohane and Victor (2011: 13) identified as a cooperation problem in the case of climate change, namely "the coordination of common scientific assessments to increase the public good of shared knowledge about the causes and consequences of [in this case, biodiversity loss]." Indeed, as emphasized in decision IPBES-2/4 (UNEP, 2014),[14] "Assessments, whether global, regional or thematic, need coherence in their approach, which will provide opportunities for synthesis between the assessments, the scaling up and down of assessments done at different scales, and also comparison among assessments performed at specific scales or on different themes."

In contrast with IPCC, in fulfilling its four core missions (catalyze the generation of new knowledge, perform assessments, develop policy support tools, and promote capacity-building), IPBES is to serve the Complex rather than one single convention. It also has more godfathers than IPCC: although administratively run by UNEP, it has formal institutional links with UNESCO, FAO, and UNDP (see below). Moreover, it must also address the needs of a set of MEAs that are at the core of the Complex (CBD, Ramsar, WHC, CITES, CMS and the International Treaty on Plant Genetic Resources for Food and Agriculture (ITPGRFA). These regimes are dominated by different knowledge and epistemic communities[15] that do not define the biodiversity problem or solutions to it in the same way, even though they may all agree with the core proposition that biodiversity is eroding and must be protected. In this setting, to what extent can IPBES operate effectively and facilitate convergence among those various knowledge communities when it is itself dominated by one of them? The first question that arises in this context pertains to the elements of the Complex that the IPBES should help coordinate. The potential size of the Complex is huge. Knigge, Herweg and Huberman (2005), for example, have identified some 155 conventions dealing with biodiversity issues at different levels – and this number refers only to intergovernmental institutions...

First come governments through IGOs. Even though specific epistemic/knowledge communities played a role in its creation, IPBES is the product of governments' preferences. Hence, it must fulfill the functions that they wish to see emerge from the Complex. Its conceptual framework, for example, must accommodate their concerns and its operations give them more control over scientific advice. Then come the scientific subsidiary bodies of the main biodiversity-related MEAs (such as the CBD's SBSTTA) and other transnational scientific bodies and programs (such as the International Council of Scientific Unions (ICSU) and GeoBon[16]). Finally come transnational civil society organizations (IUCN, Indigenous Peoples Forum).

The second question pertains to the kind of issues on which IPBES should facilitate convergence, while keeping in mind that: (i) biodiversity is about values and the relationship between human societies and the natural world, and not just about science, and that (ii) the regimes forming the Complex were created at different times and thus reflect different norms, concerns, and approaches. In this light, IPBES would focus on problem definition (through the Conceptual Framework), knowledge (through assessments), priorities (through specific policies for sorting out requests), and actions (measures intended to improve the science-policy interface). The challenge is to create an intersubjective understanding of the contents of biodiversity science in the face of different knowledge systems, competing policy priorities, and rival political and normative agendas.

Finally, the third question pertains to the means for doing so, and we will consider what has been put in place so far around the institutionalization of scientific discourse and the means to foster horizontal and vertical coordination.

The legitimization of scientific discourse

Politics begin with issue framing. Institutions created to pursue specific policies embody a particular conception of the relationship between human societies and the natural world. IPBES, therefore, aims to become the arena where the dominant discourse about biodiversity and biodiversity science is being constructed and legitimized. This discourse defines core concepts, identifies main questions, and offers a set of assumptions regarding the key relationships among them (Diaz, 2015).

The goal of IPBES is to "strengthen the science-policy interface for biodiversity and ecosystem services for the conservation and sustainable use of biodiversity, long-term human well-being and sustainable development." This formulation contains the core of the discourse: the need to make science relevant to policy and ensure that policy is informed by science; the distinction between biodiversity and ecosystem services and the (non-exclusive) emphasis on the latter; and biodiversity and ecosystem services for human well-being and sustainable development. Within the IPBES Conceptual Framework,[17] this discourse basically hinged on a socioecological system viewpoint whereby both the social and natural worlds are considered as co-constructed and where the core concerns are the complexity, resilience, robustness, sustainability, vulnerability and adaptability of the system. The main concept that links the relationship between these two dimensions (social and natural) is the ecosystem service concept, an approach that remains controversial within the scientific community itself as well as with some governmental delegations and NGOs. Finally, although nature still tends to be conceptualized as separate, social variables such as poverty alleviation and equity are prominent concerns.

The Conceptual Framework (CF) aims to promote "a shared working understanding across different disciplines, knowledge systems and stakeholders" of the Complex. It summarizes the relationships between people and nature that IPBES aims to promote and intends to help identify and prioritize knowledge gaps as well as define initiatives designed to fill them. Regarding pollination, for example, the framework would encourage experts not only to focus on the immediate causes of the loss of pollinators (spread of pathogens, habitat change, pesticides), but also on the institutions that facilitate them and increase the vulnerability of these populations, and on the impact of pollinator decline on agricultural ecosystem services. It would also encourage experts to think about solutions drawn from indigenous and local knowledge systems, their impacts on other values (poverty alleviation, equity), or to consider how to involve governments, the private sector and civil society in addressing pollinator decline.[18]

Based on the experience of the existing scientific interplay among some MEAs of the Complex, this framework indeed has the potential to influence the work of the Complex as a whole. For example, past interactions with the CBD have prompted the reformulation of key Ramsar commitments in favor of the "wise use" of wetlands instead of the "benefit of mankind" perspective in order to incorporate explicitly the ecosystem approach (Caddell 2011). Of course, this ecosystem approach was already incipient in Ramsar – such reorientation would be much more difficult in the case of species-centered conventions.

The adoption of the IPBES Conceptual Framework is an important first step, but only that. Upcoming challenges include: the extent to which the Framework will be appropriated and used, and how it will translate into legitimate knowledge; the relative emphasis given to each of its elements; how each element will be conceptualized (e.g. what questions will drive social and political analysis?); how some knowledge communities will be integrated into the Framework and how contradictions among their perspectives will be overcome (conservationists, for example, have long been critical of the ecosystem services approach); how different ontologies and epistemologies will be reconciled;[19] and how the Framework will evolve in the future.[20] Beyond the development and appropriation of a Framework that serves both levels, the coordination challenge entails specific actions with horizontal and vertical dimensions.

This Framework still needs to be operationalized, which will be achieved in practice through the assessments performed. It should, in turn, guide the four functions of IPBES mentioned previously. It is intended not only to support the analytical work of the Platform, but also to "guide the development, implementation and evolution of its work programme, and to catalyse a positive transformation in the elements and interlinkages that are the causes of detrimental changes in biodiversity and ecosystems and subsequent loss of their benefits to present and future generations."[21] This is mainly achieved through assessments.

IPBES and control over knowledge

Who defines what is biodiversity science and what is considered legitimate knowledge? IPBES represents both the product of various networks (see chapter 3 in this book) and the institutionalization of a particular perspective, i.e. the ecosystem services framework of analysis and policy. Assessments are attempts to present decision makers with credible, legitimate and salient knowledge (Reid 2004), arenas where potentially "mutual learning occurs across scales and knowledge systems" (Miller & Erickson, 2006: 302). But they are also the product of power relationships among states, IGOs, knowledge communities (such as ecological economics, conservation, biodiversity, genetics, evolution, ecology, cultural anthropology, sociology, law, political science, indigenous and local communities) and laboratories (that promote specific concepts and techniques). Thus, despite the existence of the Conceptual Framework, governments as well as competing knowledge communities are eager to take part in scoping these assessments, a process that is extensively detailed in IPBES, as well as in the interpretation of their results.

As a consequence, much of the IPBES politics are focused on the conditions governing the nomination of members of the Multidisciplinary Expert Panel (MEP) and various expert committees (including their links with governments and the disciplines represented), as well as the selection and production of its main deliverables. Who can submit assessment requests? How is IPBES to choose among them? How should they be conducted? As specified in the procedures for preparation of the Platform's deliverables,[22] inputs and suggestions received by the secretariat are considered and prioritized by the MEP and the Bureau. These requests may come from governments through MEAs, from joint submissions by multiple governments, including regional groups, and from other sources. Should the MEP and the Bureau conclude that additional scoping is required to complete the prioritization of certain requests (more of a rule than an exception), the MEP will submit a proposal to that end to the Plenary for consideration and then select experts from lists of nominations by governments and relevant stakeholders, with government nominees representing at least 80% of the experts. Following the scoping process, MEP and the Bureau decide whether to proceed with the assessment, assuming that it can be conducted within the budget and timetable approved by the Plenary. If, however, the Panel and Bureau conclude that the assessment should not go forward or diverge on that matter, they will inform the Plenary for review and decision support regarding the matter. Clearly governments remain in control of the whole process, a requirement that is intended to foster government appropriation of the results.

The coordination challenge

This challenge covers the production, reconciliation, and impact of norms and knowledge regarding biodiversity protection and its uses. The regime complexity raises an obvious problem for governments concerning policy coherence (Morin & Orsini, 2013). Beyond the fragmentation of the complex, this problem may be enhanced by the difficulty in mapping the specific policies pursued by MEAs on particular issues (Caddell, 2011). IPBES might thus represent an alternative means of coordination. If the Complex develops means of facilitating interplay, policy coherence may co-evolve as a result of interactions among actors and institutions across governance levels (Morin & Orsini, 2013). This process is not automatic. Further, the coordination issue is not as much a problem of public policy arising from the implementation of overlapping commitments (Velázquez Gomar, Stringer & Paavola, 2013), but rather – from an upstream perspective – one of coherence among various commitment definitions. Hence, even with IPBES fulfilling its coordination role, individual organizations and groups that operate within the Complex would still approach the problem differently according to their own norms, technologies, and experience, yet the problem would be defined jointly and the criteria for interpreting and accepting knowledge would be shared. This would safeguard the possibility of testing different solutions. Convergence does not entail uniform policy solutions.

The formation of a cluster of the six main Biodiversity MEAs led to the creation of a variety of norms, policies, as well as the adoption of measures that both overlap and complement one another. Coordination of scientific definitions of the problem, knowledge and activities may positively affect the implementation of international and national policies; in the latter case, enhanced coordination at the international level could help improve coordination at the national level in some countries (see below). When and under what conditions this may be the case still needs to be clarified. Velázquez Gomar, Stringer & Paavola (2013) suggest that it is far from automatic and, indeed, observers have long argued the reverse, namely that international coordination begins at home. Clearly, this may depend on national characteristics, such as the level of national wealth, the degree of conflict in society, or the nature of the state.

At the initiative of CBD, cluster members have made significant efforts to improve cooperation by sharing information, reducing competition, and aligning policies through mechanisms such as the Biodiversity Liaison Group (BLG) and the Chairs of the Scientific Advisory Bodies of Biodiversity-related Conventions (CSAB) group. MEA secretariats have also signed memorandums of understanding (MOUs) that aim at reducing overlaps and competition while fostering working relationships on matters of common concern, some of which "have subsequently spawned joint initiatives, work programs, and thematic policies, with varying degrees of success and efficiency" (Caddell, 2011). Apart from the core cluster, "collaboration beyond the biodiversity cluster involves generic (e.g. the UN Environment Management Group) and thematic (e.g. the Inter-Agency Liaison Group on Invasive Alien Species) mechanisms for inter-institutional co-operation" (Velázquez Gomar, Stringer & Paavola, 2013).

The results have been limited, however, and largely ad-hoc. MOUs and Joint Work Plans have been signed, but synergistic solutions that would apply to a variety of conventions are lacking (Urho, 2009). In this context, rather than being conceived as potentially hierarchically superior to MEAs and other elements of the complex,[23] IPBES will operate as a focal point where problem definitions are negotiated (provided that the IPBES framing discourse is accepted) and policy instruments developed and coordinated.[24]

The horizontal coordination challenge

Within the Complex, cooperative arrangements have often been undermined by resource constraints (Caddell 2011), an obstacle that IPBES might be able to overcome by minimizing the transaction costs associated with collaboration. IPBES fulfills this function primarily through its procedure for receiving and prioritizing requests put to the Platform (see above), through the development of collaborative arrangements with its parent organizations (UNESCO, FAO, UNDP, UNEP),[25] and through strategic partnerships with selected organizations (such as the CBD).

The core circle of partners involves four organizations that have striven to play a significant role in the Platform when its structure was negotiated. On the basis of a revised joint proposal by the four UN organizations, the first Plenary requested UNEP, UNESCO, FAO and UNDP to establish an institutional link with the Platform through a collaborative partnership arrangement for the work of IPBES and its secretariat.[26] Consequently, these "partners" have a privileged status that comes close to joint administration of the Platform (although UNEP, as *primus inter pares*, plays a leading role and is providing the Platform secretariat). It involves staff seconded to the Platform and an internal dedicated capacity to support it; collaboration in the recruitment of the head of the Platform secretariat and of other professional posts in the secretariat; technical support; joint fundraising; delegation of special tasks to one or more partners;[27] exchange of information, including consultation prior to the publication of Platform documents (including official meeting documents which bear the logos of the four partners); and the right to participate in meetings of the Platform Plenary and of the Multidisciplinary Expert Panel and other subsidiary bodies to be established by the Plenary.[28] Resources dedicated to IPBES by these organizations remain limited, however.[29]

In 2015, the Plenary adopted a decision that sets out policy guidelines for entering into partnerships with multilateral environmental agreements and academic, scientific and United Nations system organizations, with UN bodies and MEAs being given priority at this stage.[30] The model used was a memorandum of understanding signed in October 2014 between the Platform secretariat and the CBD secretariat.[31] Such MOUs subsequently have to be approved by the Plenary.

Yet much remains to be done regarding enhancing collaboration within the Complex. The chairs of the MEAs' scientific subsidiary bodies (CSAB) are not yet routinely allowed to participate as observers in MEP meetings.[32] Regarding the Biodiversity Liaison Group (BLG), IPBES remains a forum where these IGO members should discuss further means of cooperation, especially with respect to knowledge management and the exchange of shared models of effective policy.[33] Clearly also CSAB should become a key mechanism in fulfilling this coordination function as it allows a variety of regional actors, UN bodies, and NGOs to participate in its meetings. In 2013 (CSAB-6), participants agreed that, in addition to a member of the IPBES Secretariat, an IPBES representative, e.g. its chair or a co-chair of MEP, would be invited to future CSAB meetings.

CSAB has mainly focused on "examining areas of cooperation and translating scientific considerations into clear policies, alongside the identification of emerging issues with reference to problems and priorities within the individual scientific fora of the participating regimes" (Caddell, 2011: 61). Two key collaborative policies have recently emerged. First, "considerable attention has been accorded to implementing strategic plans in a more integrated and coherent manner as well as to providing scientific leadership for the further development of National Biodiversity Strategies and Action Plans (NBSAPs) mandated under CBD. Second, following the difficulties identified by CITES and CMS over scientific inconsistencies between regimes, "the harmonization of species nomenclature is to be addressed as a standing item" (Caddell, 2011: 62). Its action could therefore be key in linking global and local levels.

The vertical coordination challenge: towards reconciling global and local levels

Although a regime complex involves various levels of analysis, the properties of one level are not automatically extended to other levels – what is true at the global level may not be so at the regional and local levels, and conversely. Linking the global, regional, sub-national and local levels of analysis and action remains a challenge, in part to avoid inconsistencies between them, and also because fragmentation at lower levels will encourage fragmentation at the global level. The aim is to allow for the development of an intersubjective understanding among those levels of governance, one able to support the mobilization of stakeholders (actors taking part in the definition and implementation of regimes). This is all the more important since the existing global-local links between elements of the Complex and national and local units may encourage the fragmentation of national biodiversity policies (Le Prestre, 2010).

Bernstein & Cashore (2012) identify four different pathways through which global governance arrangements can influence national policy, two of which, falling under the IPBES mission, are: (i) international norms and discourse setting general standards of behavior of non-binding character, and (ii) direct access to domestic-policy making processes in the form of education, training, assistance, capacity-building and/or co-governance via partnerships.

At this stage of its development though, not much can be said regarding the role that IPBES actually plays in reconciling the global, regional, and local levels of analysis and actions. Since it is to operate at the global or regional level only, it has not developed specific actions designed to address this link. Yet it is still not clear to what extent global coherence will enhance local coherence, not only among biodiversity-related activities undertaken by different ministries, but also among different sectors such as environment, development and agriculture. This is unlikely without sustained action. Several global organizations, such as UNEP, UNDP, IUCN and other IGOs "support domestic efforts to enhance integration of biodiversity-related MEAs" (see Andresen & Rosendal, 2009; Urho, 2009). Yet, in Latin America, national coordination of implementation activities has lagged behind global synergies in the biodiversity cluster. "Feedback loops between governance levels have not been strong enough to bridge that gap" (Velázquez Gomar, Stringer & Paavola, 2013: 1).

Conclusion

A regime complex must be managed in order to deliver (Keohane & Victor, 2011). By playing a key scientific coordination role, IPBES could help overcome some of the negative consequences of a fragmented policy field, namely forum-shopping, regime-shifting, strategic inconsistency (Alter & Meunier, 2009) or inconsistencies in problem-definitions and resulting policies.

The effectiveness of IPBES in acting as the scientific coordination arena (with an impact on the nature of policy options open to the Complex components) remains open, however. Certainly, it has so far put considerable emphasis on devising procedures intended to foster credibility, relevance, legitimacy, and transparency in response to the challenges of implementing a non-linear model of the relationship between science and policy (Koetz, Farrell & Bridgewater, 2011). Beyond the effectiveness of the procedures it has put in place to that end, many questions arise regarding the appropriation and operationalization of the Conceptual Framework, the integration of complex systems properties into its operationalization, the impact of the Platform on the scientific responsibilities of individual regimes and on the construction and consolidation of knowledge systems, and about the ability of IPBES to act as a learning and adaptive device for the whole Complex through its assessments, its impact on the strategies of the MEAs scientific bodies, and the evolution of its Conceptual Framework.

Indeed, although the latter (drawing lessons from the Millennium Assessment) emphasizes the importance of feedbacks and links between the social and natural, much remains to be done to integrate more fully features of complexity theory into the work of IPBES. Socioecological systems are complicated and should be approached as such. Consequently, it will be particularly important in future assessments to include uncertainty into scientific models, consider unlikely events, expand the integration between social and ecological systems (i.e. not define the social too narrowly but also take structural changes regarding values into account, for example), integrate more social feedback, and identify tipping points and risks of non-linear changes (see Wells, 2009).

The political challenge of fulfilling this ambition remains daunting. First, science bears interests and power. Different international regimes embody different approaches to the biodiversity issue and are linked to competing scientific communities. Second, IPBES runs the risk of being captured by a particular knowledge community or international organization, of having to dilute its mandate by trying to accommodate all stakeholders within the Complex, or of seeing its efforts to promote cooperation and convergence blocked by organizations and groups fearful of losing influence and autonomy. Various stakeholders will be keen to use IPBES to fulfill their own political goals. And third, governments have multiple opportunities at their disposal, besides IPBES, to influence the process and deliverables in order to promote their own definition of the problems, their notion of what deserves to be measured, and the best solutions to address these problems.

Notes

1 Its current members are: the Convention on Biological Diversity, the Convention on the Conservation of Migratory Species of Wild Animals, the Convention on International Trade in Endangered Species of Wild Fauna and Flora, the International Plant Protection Convention, the International Treaty on Plant Genetic Resources for Food and Agriculture, the Ramsar Convention on Wetlands, and the World Heritage Convention.
2 A framework convention is a legally binding international treaty establishing general guidelines and principles for international governance on a particular issue. Separate and more detailed legal instruments called protocols can be attached to a framework convention to address specific aspects of an issue.
3 By which the CBD "acknowledges the significance of traditional knowledge and practices, which should be taken into account in the implementation of all aspects of the Convention" and needs to be safeguarded by the parties, according to the CBD website (www.cbd.int/gbo1/chap-02.shtml).
4 Trade-related aspects of intellectual property rights (TRIPS).
5 Food and Agriculture Organization of the United Nations (FAO) and World Intellectual Property Organization (WIPO).
6 According to the authors, a regime is "an international institution, based on an explicit agreement, that reflects agreed principles and norms and codifies specific rules and decision-making procedures". It is slightly different from the iconic definition of regimes in Krasner 1983.
7 Although the G77 countries prevented the adoption of the proposed MEA against deforestation in 1992, and insisted that forest management was a matter of national sovereignty, the rapid erosion of rainforests is a key issue as part of the broader biodiversity erosion issue. Rainforests contain by far the largest and least studied part of global biodiversity.
8 TRAFFIC, created in 1976 with WWF, monitors the illegal trade of wild flora and fauna species. With UNEP and WWF, IUCN also manages the World Conservation Monitoring Centre established in 1988.
9 The IUCN "Red List of Threatened Species" and Conservation International's favored concept of "biodiversity hotspots", adopted by the organization just a

year after biologist Norman Mayers introduced the concept (see www.conserva tion.org/How/Pages/Hotspots.aspx), are cases in point.

10 The biologist Walter G. Rosen, from the US National Research Council, allegedly suggested the word in 1985, during the first preparatory meeting of the US National Forum on Biodiversity.

11 Out of the six conventions in the cluster, only the Convention Concerning the Protection of the World Cultural and Natural Heritage does not have a dedicated scientific and technical advisory body, as it relies on other organizations for scientific advice.

12 According to the CBD website, "Multidisciplinary and open to participation by all Parties, SBSTTA comprises government representatives competent in the relevant field of expertise" and meets every year (www.cbd.int/sbstta).

13 A central character in this group is Robert T. Watson, Chairperson of the IPCC from 1996 to 2002, also Chair of the GBA, Co-chair of the Millennium Assessment, Chair of the GEF Scientific and Technical Advisory Panel from 1991 to 1994 and Director of the International Assessment of Agricultural Knowledge, Science and Technology for Development (IAASTD). He played a pivotal role in the diffusion of the ecosystem service concept. He is currently a member of the IPBES Bureau.

14 "Decision IPBES-2/4: Conceptual Framework for the Intergovernmental Science-Policy Platform on Biodiversity and Ecosystem Services," in UNEP (2014), *Report of the second session of the Plenary of the Intergovernmental Science-Policy Platform on Biodiversity and Ecosystem Services*, Nairobi: UNEP (Document IPBES/2/17).

15 Whereas knowledge communities are simply composed of experts/scientists that share a common paradigm, epistemic communities, following Haas' seminal work (Haas 1990), translate this shared knowledge into policy preferences that they undertake to promote by being closely connected with government institutions and decision-making circles.

16 The Group on Earth Observations Biodiversity Observation Network (GeoBon).

17 Decision IPBES/2/4 and documents IPBES/2/INF/2 and Add.1 (2013).

18 It is apparent in this presentation that the IPBES Conceptual Framework (or, rather, Multidisciplinary Expert Panel members) has yet to fully integrate the non-linear model of science-policy relationship that IPBES is supposed to embody. Indeed, one of the first questions that one would ask with regard to such a model would be: How do various stakeholders define the problem? Rather, the presentation of a problem requiring an assessment, such as pollinator decline, starts with the identification of a trend by scientists that is defined as an issue that must be addressed (How to reverse it?).

19 Indigenous knowledge should not be seen as complementary (as indicated in the scoping of the pollinator decline assessment) but as an alternative vision. Likewise, the function of social sciences is not just to support the definition of the problem as expressed by natural scientists or economists. One way to reconcile these ontologies and epistemologies is through the institutionalization of dedicated processes, such as the Panama dialog workshop.

20 Adopted by a decision, it will have to be amended by a decision of the IPBES Plenary.

21 "Conceptual Framework for the Intergovernmental Science-Policy Platform on Biodiversity and Ecosystem Services," in UNEP (2014), *Report of the second session of the Plenary of the Intergovernmental Science-Policy Platform on Biodiversity and Ecosystem Services*, Nairobi: UNEP (Document IPBES/2/17), p.39.

22 Document IPBES/3/18 and decision IPBES/1/3 (Procedure for receiving and prioritizing requests put to the Platform).

23 This approach thus differs from a more centralized conception where, as expressed by the French government when it first proposed the creation of a scientific biodiversity panel on the model of IPCC that later became IPBES, it would be in the driver's seat and capable of imposing directives.

24 See, for example, the work of the IPBES Expert Group in charge of developing a guide on and a catalogue of policy support tools and methodologies (deliverable 4c).
25 Docs: IPBES/2/15, IPBES/2/INF/3, IPBES/3/INF/14; decision IPBES/2–8.
26 Decision IPBES/1/4.
27 For example, UNESCO is taking the lead on indigenous and local knowledge as well as knowledge data, whereas UNDP is leading on capacity building.
28 Document IPBES/2/15.
29 UNESCO hosts the technical support unit (TSU) for the task force on indigenous and local knowledge systems (ILK-TF) with one full-time equivalent of UNESCO staff. UNDP was requested through the Busan Outcome in 2010 to play a special role in developing capacity to support the Platform, integrating capacity building with assessment, knowledge generation, and policy-relevant tools and methodologies to help countries tackle science-policy questions critical to sustainable development. BES-Net, the Biodiversity and Ecosystem Services Network, is UNDP's response to this request.
30 Paragraphs 6–8 of decision IPBES-3/4. See also the draft decision on strategic partnerships (IPBES/3/L.8), as well as IPBES/2/14; IPBES/3/17.
31 See decision IPBES-3/4, Annex III: Guidance on the development of strategic partnerships and other collaborative arrangements. (doc. IPBES 3/17).
32 See, for example, decision CBD/SBSTTA/18/9 (2014) requesting the CBD Executive Secretary to facilitate such participation.
33 At their ninth meeting (2014), BLG members agreed that the IPBES Secretariat should be invited to the next BLG meeting and that there would be a dedicated agenda item on IPBES cooperation, focusing especially on scientific and technological cooperation and the agreed thematic assessment on sustainable use.

References

Abbott, K. W. 2011. "The Transnational Regime Complex for Climate Change." *Environment & Planning C: Government & Policy*, 30, 571–590.

Alter, K. J. & Meunier, S. 2009. "The Politics of International Regime Complexity." *Perspectives on Politics*, 7, 13–24.

Andersen, R. 2008. *Governing Agrobiodiversity: Plant Genetics and Developing Countries.* Aldershot: Ashgate.

Andresen, S. & Rosendal, G. K. 2009. "The Role of the United Nations Environment Programme in the Coordination of Multilateral Environmental Agreements." In: Biermann, F., Siebenhüner, B. & Schreyoegg, A. (eds.) *International Organizations and Global Environmental Governance.* London: Routledge.

Bauer, S. 2009. "The Secretariat of the United Nations Environmental Programme: Tangled Up in Blue." In: Biermann, F. & Siebenhüner, B. (eds.) *Managers of Global Change: The Influence of International Environmental Bureaucracies.* Cambridge, MA: MIT Press, 169–202.

Beck, S. et al. 2014. "Towards a Reflexive Turn in the Governance of Global Environmental Expertise. The Cases of the IPCC and the IPBES." *GAIA- Ecological Perspectives for Science and Society*, 23, 80–87.

Bernstein, S. & Cashore, B. 2012. "Complex Global Governance and Domestic Policies: Four Pathways of Influence." *International Affairs*, 88, 585–604.

Biermann, F., Pattberg, P. & Van Asselt, H. 2009. "The Fragmentation of Global Governance Architectures: A Framework for Analysis." *Global Environmental Politics*, 9, 14–40.

Caddell, R. 2011. "The Integration of Multilateral Environmental Agreements: Lessons from the Biodiversity-Related Conventions." *Yearbook of International Environmental Law*, 22, 37–75.

Diaz, S. E. A. 2015. "The IPBES Conceptual Framework – Connecting Nature and People." *Current Opinion in Environmental Sustainability*, 14, 1–16.

Haas, P. M. 1990. *Saving the Mediterranean: The Politics of International Environmental Cooperation.* New York: Columbia University Press.

Jardin, M. 2010. "Global Biodiversity Governance: The Contribution of the Main Biodiversity-Related Conventions." In: Billé, R.Chason, L., Chiarolla, C., Jardin, M., Kleitz, G., Le Duc, J.-P.. & Mermet, L. (eds.) *Global Governance of Biodiversity: New Perspectives on a Shared Challenge.* Paris: IFRI, 6–44.

Keohane, R. O. & Victor, D. G. 2011. "The Regime Complex for Climate Change." *Perspectives on Politics*, 9, 7–23.

Knigge, M., Herweg, J. & Huberman, J. 2005. "Geographical Aspects of International Environmental Governance, Illustrating Decentralisation." Berlin: Ecologic Institute for International and Environmental Policy.

Koetz, T., Bridgewater, P., Van den Hove, S. & Siebenhüner, B. 2008. "The Role of the Subsidiary Body on Scientific, Technical and Technological Advice to the Convention on Biological Diversity as Science-policy Interface." *Environmental Science and Policy*, 11, 505–516.

Koetz, T., Farrell, K. N. & Bridgewater, P. 2011. "Building Better Science-policy Interfaces for International Environmental Governance: Assessing Potential within the Intergovernmental Platform for Biodiversity and Ecosystem Services." *International Environmental Agreements: Politics, Law and Economics*, 12, 1, 1–21.

Krasner, S. D. (ed.) 1983. *International Regimes.* Ithaca, NY: Cornell University Press.

Larigauderie, A. & Mooney, H. A. 2010. "The Intergovernmental Science-Policy Platform on Biodiversity and Ecosystem Services: Moving a Step Closer to an IPCC-Like Mechanism for Biodiversity." *Current Opinion in Environmental Sustainability*, 2, 9–14.

Le Prestre, P. 2002. "The Operation of the Convention Governance System." In: Le Prestre, P. (ed.) *Governing Global Biodiversity.* Aldershot, UK: Ashgate.

Le Prestre, P. 2010. "La nécessité d'une gouvernance interscalaire de la biodiversité." In: Nègre, C. (ed.) *La Convention internationale sur la biodiversité: enjeux de la mise en œuvre.* Paris: La Documentation française.

Le Prestre, P. & Martimort-Asso, B. 2004. "Les questions soulevées par le système de gouvernance internationale de l'environnement." Paris: IDDRI.

Miller, C. & Erickson, P. 2006. "The Politics of Bridging Scales and Epistemologies Science and Democracy in Global Environmental Governance." In: Reid, W. V., Berkes, F., Wilbanks, T. & Capistrano, D. (eds.) *Bridging Scales and Knowledge Systems Concepts and Applications in Ecosystem Assessment.* New York: Island Press.

Morin, J.-F. & Orsini, A. 2013. "Regime Complexity and Policy Coherency: Introducing a Co-Adjustments Model." *Global Governance*, 19, 41–51.

Oberthür, S. 2002. "Clustering of Multilateral Environmental Agreements: Potentials and Limitations." *International Environmental Agreements: Politics, Law and Economics*, 2, 317–340.

Oberthür, S. 2009. "Interplay Management: Enhancing Environmental Policy Integration Among International Institutions." *International Environmental Agreements: Politics, Law and Economics*, 9, 371–391.

Oberthür, S. & Gehring, T. (eds.) 2006. *Institutional Interaction in Global Environmental Governance. Synergy and Conflict among International and EU Policies*. Cambridge, MA: MIT Press.

Oberthür, S. & Stokke, O. S. (eds.) 2011. *Managing Institutional Complexity: Regime Interplay and Global Environmental Change*. Cambridge, MA: MIT Press.

Orsini, A. & Compagnon, D. 2013. "Les acteurs non étatiques dans les négociations multilatérales." In: Petiteville, F. & Placidi-Frot, D. (eds.) *Négociations Internationales*. Paris: Les Presses de Sciences Po.

Orsini, A., Morin, J.-F. & Young, O. 2013. "Regime Complexes: A Buzz, a Boom, or a Boost for Global Governance?" *Global Governance*, 19, 27–39.

Raustiala, K. & Victor, D. G. 2004. "The Regime Complex for Plant Genetic Resources." *International Organization*, 58, 277–309.

Reid, W. V. 2004. "Bridging Scales and Epistemologies in the Millennium Ecosystem Assessment." Paper given at the Conference on *Bridging Scales and Epistemologies – Linking Local Knowledge and Global Science in Multi-Scale Assessments*. Alexandria, Egypt.

Rosendal, K. G. 2006. "The Convention on Biological Diversity: Tensions with the WTO TRIPS Agreement over Access to Genetic Resources and the Sharing of Benefits." In Oberthür, S. & Gehring, T. (eds.), *Institutional Interaction: Enhancing Cooperation and Preventing Conflicts between International and European Environmental Institutions*. Cambridge, MA: MIT Press, 79–102.

UNEP 2014. *Report of the Second Session of the Plenary of the Intergovernmental Science-Policy Platform on Biodiversity and Ecosystem Services*. Nairobi: UNEP (Document IPBES/2/17).

Urho, N. 2009. *"Possibilities of Enhancing Cooperation and Co-Ordination among MEAs in the Biodiversity Cluster."* Copenhagen: Nordic Council of Ministers.

Velázquez Gomar, J. O. 2014. "Environmental Policy Integration among Multilateral Environmental Agreements: the Case of Biodiversity." *International Environmental Agreements: Politics, Law and Economics*, 1–17.

Velázquez Gomar, J. O., Stringer, L. C. & Paavola, J. 2013. *"Regime Complexes and National Policy Coherence: Experiences in the Biodiversity Cluster."* Leeds, UK: University of Leeds, Sustainability Research Institute (SRI), School of Earth and Environment.

Wells, J. L. 2009. "Complexity and Climate Change: An Epistemological Study of Transdiciplinary Complexity Theories and their Contribution to Socio-ecological Phenomena." Ph.D. Dissertation, University of California, Berkeley.

Young, O. R. (ed.) 2002. *The Institutional Dimensions of Environmental Change: Fit, Interplay, and Scale*. Cambridge, MA: MIT Press.

Zelli, F. & Van Asselt, H. 2013. "Institutional Fragmentation of Global Environmental Governance: Causes, Consequences, Responses." *Global Environmental Politics*, 13, 1–13

3   The birth of a science-policy interface for biodiversity

The history of IPBES

Alice B.M. Vadrot

Introduction – IPBES as historical phenomenon

The creation of the Intergovernmental Platform on Biodiversity and Ecosystem Services (IPBES) in April 2012 can be seen as a milestone in international biodiversity politics, policies and science. The new institution is expected to tie together existing (scientific) knowledge on different aspects of the Earth's biological diversity, based on common methodologies and tools. This should provide decision makers at different levels with policy relevant knowledge. Furthermore, the new institution was conceptualized as a service provider, aiming at delivering tangible products (e.g. global and regional assessments) in order to raise awareness of the negative consequences of biodiversity loss, the value of nature for human wellbeing and the necessity to support and invest in natural resource conservation activities around the globe among policy makers and the public at large.

At a first glance, the establishment of a new body in the institutional landscape of biodiversity politics and science looks like a success story that is beneficial for all actors involved. It follows the principles of evidence-based politics, multilateralism and deliberative democracy and contributes to the objectives of credibility, salience and legitimacy as an integral part of linking knowledge and action for environmental decision making (Cash et al., 2003). Since its emergence in the early 1990s, international biodiversity politics in general and the Convention on Biological Diversity (CBD) in particular were confronted with a lack of (scientific) knowledge on how to tackle biodiversity loss. This had *inter alia* triggered the establishment of the CBD Subsidiary Body on Scientific Technical and Technological Advice (SBSTTA) and led to the elaboration of the Global Biodiversity Assessment (GBA) (Di Castri & Younès, 1996; Guay, 2002; McConnell, 1996). The nonavailability of scientific evidence and usable knowledge was one side of the coin. The other side of the coin regards the failure of CBD "to institutionalize the common responsibility of humanity to protect biodiversity" (Guruswamy, 1998: 351).

In this regard, the establishment of IPBES addresses some of the needs identified throughout the debate on reforming the institutional architecture of international environmental politics (Biermann and Bauer, 2005; Kanie et al., 2012). In 2005, the 60th session of the UN General Assembly recognized "the need for more efficient activities in the UN system, with enhanced coordination, improved policy advice and guidance, strengthened scientific knowledge, assessment and cooperation" (UNGA, 2005: 37). Seven years later, the United Nations Conference on Sustainable Development, Rio +20, seems to have confirmed that "[t]he practice of multilateralism has simply not caught up with structural changes in the system" (Bernstein, 2013: 16). This is particularly true for biodiversity politics and CBD: "In the absence of a common natural resource, what is it intended to regulate? What is it supposed to achieve?" (Swanson, 1999: 308). Furthermore, "widely varying concepts of nature meet (depending on the viewpoints on ecosystems, species or genetic resources; from untouched nature or the 'natural wealth of the tropics' to the utility of genetic resources), but also widely varying societal nature relations (above all diverging forms of use)" (Görg & Brand, 2000: 378). In the light of the multiple contestations of multilateral environmental politics and the contested and multifaceted "object to be governed", the establishment of IPBES could be considered a success story.

What Désirée McGraw assumed for the emergence of CBD, namely that it "entered a legal field crowded with agreements" is true for IPBES. International biodiversity policy is a rather fuzzy field, overlapping with issues of climate change, the destruction of local livelihoods, poverty reduction, food and water security; the regulation of the use of resources for the development of seeds, drugs and cosmetics, and related issues of intellectual property rights; biotechnology; and access to, and benefit sharing of genetic resources (Le Prestre, 2002; McGraw, 2002; Rosendal, 2001; Vadrot 2011; Vadrot, 2014). The complex institutional landscape of international biodiversity politics within which IPBES is operating is increasingly being seen as deficient, fragmented, and unstructured (Le Prestre, 2002). This mirrors the ineffectiveness of the current system in dealing with the ecological crisis politically, institutionally and scientifically (Görg & Brand, 2000; Koetz, Farrell and Bridgewater, 2012; Vadrot, Heumesser & Ritzberger, 2010). Since the establishment of the Convention on Biological Diversity (CBD) and its Subsidiary Body on Scientific Technical and Technological Advice (SBSTTA), the politicized character of biodiversity knowledge became a central feature of biodiversity politics and policies (Brand, 2010a, 2010b; Brand & Vadrot, 2013; Koetz et al., 2008).

In this light, the establishment of IPBES is not a matter of course. Since the early beginnings of the debate on whether to establish and institutionalize a science-policy interface for biodiversity, the dialogue between different stakeholders has been politicized. Due to politicization of the issue, the outcome of the multi-stakeholder meetings under the auspices of the United Nations Environment Programme (UNEP) from 2008 to 2010 and of the first two plenary meetings in 2011 and 2012 were for a long time unpredictable: different actors had divergent interests, problem understandings and expectations towards a new interfacing institution between science and policy for biodiversity. This has led to many predictions that the attempt to establish a science-policy interface will fail. Why then has the formal establishment of IPBES succeeded in the end? An answer to this question necessitates both empirical work and theoretical reflection on policy change as a historical development.

The history of IPBES and the policy change concept

Against this background, the history of IPBES has to be understood in relation to the challenges of coping with policy dynamics, especially since the ontology underlying the analysis is not a positivist one. Instead of conceptualizing change as observable and describable differences between a series of snapshots taken at different points in time, this analysis focuses on spatiotemporal patterns characterizing the constitution of IPBES and systemizes them in terms of narratives and epistemic selectivities (Brand & Vadrot, 2013; Vadrot, 2014a). The epistemic selectivity concept rests upon the assumption that selectivity patterns lead to the domination of specific knowledge, problem perceptions, and narratives over others and that these take part in the stabilization of how the object to be governed is understood and assessed. An important step towards stabilization is the manifestation of such knowledge, perceptions and narratives in the form of institutions, and in the shape of what people expect from the establishment of specific institutions such as IPBES.

One precondition for this endeavour is the assumption that institutions, instead of being reified, need to be reconstructed and the underlying strategies, struggles, tactics and interests need to be analysed in a wider strategic-relational context. The advantage of this concept is that institutionalization processes are not only perceived as involving the conduct of agents and their conditions for action, but also "the very constitution of agents, identities, interests, and strategies" and related competing views of how the institution should develop (Jessop, 2004: 46). In short, understanding agency and structure as dialectically co-evolving and mutually selectively structured allows us to systematically grasp the interdependencies between institutional and epistemic change. The (re-)arrangement and transformation of institutions has an impact on the formation of the actors' interests and "on the way they act in relation to their interests" and expectations about the future (Agrawal, 2005: 98). In order to be heard within a highly controversial, politicized, but at the same time rigidly regulated and structured setting of international negotiations, such expectations need to be expressed in a way that reproduces the selectively structured institutional, discursive and epistemic context. These processes form the mechanisms through which narratives emerge within a particular spatiotemporal constellation.

Structure and contents of the chapter

In what follows I will reconstruct the IPBES history and the process that led to its founding by looking at the different periods of its establishment. This chapter is the first attempt to provide a short and condensed version of what could be described as the conflicts around the emergence of IPBES as a new institution (Vadrot, 2014a).

On the basis of particular events leading to the launching of IPBES, the first and main part of this chapter is devoted to the reconstruction of the policy-making process. I put the focus on the formal institutionalization and on the results of negotiations at the following international meetings: the three multi-stakeholder meetings under the auspices of UNEP, the two plenary sessions held to determine the functions and the role of IPBES, and several meetings of the Conference of the Parties to the Convention on Biological Diversity (CBD) over the 2008 to 2012 period. The second part describes three types of narratives linking them back to the complex emergent phenomena that characterize international biodiversity politics and policy today. In doing so, I intend to make sense of the many different events shaping the way in which the institutionalization of a science-policy interface for biodiversity became successful – as a result of continuous struggle and debate on how to tackle biodiversity loss and the respective knowledge production and use.

Early beginnings: "IPCC for biodiversity"

> Well I don't think that it has emerged now. It has got a higher profile now. It has always been an issue. If you go right back to the very beginning, there was a meeting in Mexico in 1994 where 10 parties of the convention supposed scientific representatives to Mexico to propose which science input and which scientific base the convention should have. And they agreed in the way that the SBSTTA should be such a body, how it should work and how it could work. But the next year when they had the very first meeting it was actually quite difficult, because some of the key countries did not send scientists, but high-level diplomats.
>
> (European Platform for Biodiversity Research Strategy (EPBRS),
> UK Joint Nature Conservation Committee Head of Unit,
> Interview, September 2009 in Visby/Sweden at the meeting
> of the European Platform for Biodiversity Research Strategy)

After 2005, attempts intensified to create a formalized science-policy interface for biodiversity. However, as the introductory quote suggests, the idea of interfacing science and policy for biodiversity in a more systematic manner, including the proposal to create a new intergovernmental body dates back to the beginnings of the Convention on Biological Diversity (CBD).[1] I even argue that it was present since the emergence of the term "biodiversity" itself.

'Diversity without representation'[2]

At the National Forum on BioDiversity, held in Washington DC, September 21–24, 1986, more than 60 scientists, including renowned biologists such as

Edward O. Wilson, Paul R. Ehrlich and Harold A. Mooney, discussed the need to assemble knowledge on the state of global biological diversity and the necessity to develop an international political framework for regulating the loss and the sustainable and equitable use of biodiversity.[3] In the book edited by Edward O. Wilson, he holds that the conference coincided with a perceivable rise in interest in biodiversity related issues due to two more or less independent developments: the availability of a sufficient amount of data on species extinction, deforestation and tropical biology, and the awareness of the interrelation between the conservation of biological diversity and economic development (Wilson, 1988: vi).

The debate was on the increasing awareness of scientists about the value of biodiversity knowledge and its Janus-faced character – it was a knowledge base and a potential commodity at the same time. The debate shows that scientists were becoming increasingly aware of their role in the development of markets for biodiversity products that could be used for many purposes, such as for the production of pharmaceuticals, new foods, for the substitution of fossil fuels, for fibres and for other industrial goods emerging from untapped reservoirs of the tropics (Wilson, 1988: vi). This in turn created the necessity for development and implementation of global regulatory frameworks to protect the natural environment from resource overuse. Wilson points to the specific role scientists should play in biodiversity politics and policies, *inter alia* in raising awareness of the implications of deforestation, species extinction and the application of new scientific tools such as biotechnology (ibid.). There has been a significant increase in articles on biodiversity since Wilson's book was published (Harper and Hawksworth, 1994). "Some of them [were] very relevant to comprehend the emerging issues of biodiversity; others [were] confusing this new approach with the traditional field of species diversity and richness" (Di Castri & Younès, 1996: xii). At the beginning of the 1990s, 14 different definitions of biodiversity were circulating (Jutro, 1993).[4] As a response to the unsystematic increase in biodiversity research across different scientific disciplines, the scientific community focused substantial effort on defining biodiversity and institutionalizing existing networks of relevant research and science.

In 1991, the International Union of Biological Sciences (IUBS) with the support of the Scientific Committee on Problems of the Environment of ICSU[5] (SCOPE) and UNESCO launched DIVERSITAS, a cooperative scientific programme on biodiversity. "From its very inception, the DIVERSITAS programme has been designed and launched as a multi-level and multi-scale endeavour (Di Castri and Younès, 1996; Solbrig, 1991), looking at the intersections and the integration of the different disciplines involved. It intends to provide, therefore, a *leitmotiv* and an unifying principle to all the biological world" (Di Castri & Younès 1996: 2). Di Castri and Younès argued that "the concept of biodiversity [...] derived from DIVERSITAS can serve as a 'broker' towards new science, a new development and a new society" (Di Castri & Younès, 1996: 5).

At the beginning of the 1990s, and even more so after the Convention on Biological Diversity entered into force in 1993, the framing of this "new science" became an important but contested endeavour, *inter alia* because it did not exist before as a global environmental issue. Harold Mooney, a renowned conservation biologist, described how biological diversity evolved slowly as a research question, arguing that it needed to be addressed internationally in order to be recognized as an ecological problem. In parallel, ecological economists around Charles Perrings started investigating the scope for a new research programme on the economics of biodiversity loss.

> What are the costs to society of both the local and the global deletion of species? What can be done about it? [...] what is the nature of biodiversity, and why is it ecologically significant? Others ask what is the incidence of the economic costs of biodiversity loss, and what does this mean for options open to decision makers?
>
> (Perrings, 1995: 1)

SBSTTA, which was established under CBD Article 25, was meant to at least "provide assessments of the state of scientific and technical knowledge, and the effects of various types of measures taken through efficient and state-of-the-art technologies for the conservation and sustainable use of biological diversity" (Olembo, 1994). The major challenge at that time was the lack of knowledge on what was actually known about "the extent and diversity of the world's biological resources" (McConnell, 1996: 148). In support of SBSTTA's work, the United Nations Environment Programme commissioned the Global Biodiversity Assessment, "an independent, critical, peer-reviewed scientific analysis" spanning from 1992 to 1995 (Olembo, 1994). GBA involved more than 300 scientists from over 40 countries. Edited by Vernon H. Heywood and chaired by Robert Watson, the report was conceptualized as a source of information for policy makers and practitioners on the state of knowledge on biological diversity and existing knowledge gaps (Heywood, 1995). It has emphasized uncertainties as well as issues on which scientists have reached a consensus "in the broad field of biodiversity" (Heywood, 1995, preface). However, SBSTTA was reluctant to systematically use or endorse GBA (Le Prestre, 2002: 106).

One of the GBA authors points to the challenges the authors were facing in convincing the policy makers they attempted to address: "We did a beautiful book on biodiversity and we gave it to the governments and they said: 'Well... No.' They wouldn't even look at them. Because they didn't ask for them. That didn't work."[6] According to Luis Guay, the GBA "reads like an extensive and up-to-date course in biology" (Guay, 2002: 224) lacking any presence of numerical modelling, which characterizes IPCC reports and would have been important to "produce an overall understanding of a global process" (Guay, 2002: 225). According to Frank Biermann, GBA had very limited impact on developing countries, because their knowledge needs were not sufficiently recognized and respective experts not involved (Biermann, 2002). From his perspective, GBA was "failing" because of both the way it was produced and the way its contents were chosen and presented. However, the history of GBA also shows how tensions between the different scientific communities and the divergent interests of different states and the difficulty of producing policy-relevant information are also related to the different problem understandings and expectations of policy makers and scientists and the way in which "biodiversity" is understood and addressed within the CBD framework:

> The Convention [on biological diversity] says that biodiversity is the measure of the diversity of life from genes to ecosystems. There is no way to measure that. How do you count it? They say we have the 2010 target and we are going to reduce the rate of loss of biodiversity. But how do you know if you are losing or not? For me biodiversity was a political term. There is a way to get beyond where we were in conservation, which was giant pandas, elephants, tigers. It was focusing mostly on species. What we wanted to do, really had to do with emerging technology (especially biotechnology) and genetic resources. It was how to have a convention that includes genetic resources among the species and the ecosystems that support them. So it is a concept. It is not something you can count.
>
> (International Union for the Conservation of Nature [IUCN], Head of Ecosystem Management Programme, Interview, June 2010 in Busan/Rep. of Korea at the third IPBES meeting)

In the second book on biodiversity edited by Wilson, he underlines that

> the term and concept has been a remarkable event in recent cultural evolution: 10 years ago the word did not exist, except perhaps through occasional idiosyncratic use. Today it is one of the most commonly used expressions in the biological sciences and subsequently has become a household word.
>
> (Reaka-Kudla et al., 1997: 1)[7]

In his contribution to the edited book, Thomas E. Lovejoy, Counsellor to the Secretary for Biodiversity and Environmental Affairs at the Smithsonian Institution, tries to explain what biodiversity is, arguing that "[b]iodiversity has to be thought of in a number of different ways" (Lovejoy, 1997: 7). Similar expressions can be found in a broad range of literature from various disciplines of natural and social sciences (Amano & Sutherland, 2013; Brand, 2010b; Norton & Hannon, 1997; Takacs, 1996; Vadrot, Heumesser & Ritzberger, 2010).

Hence, biodiversity did not just emerge as a new term but, since its genesis, its importance has increased at the interface between science, politics and policy. As a result, specific institutional and discursive patterns of governing biodiversity through multilateral environmental agreements emerged. The difficulty in negotiating issues related to biodiversity at the global level can be seen in both the CBD implementation deficit and the struggle of CBD-SBSTTA to provide policy-relevant science advice to the CBD Conference of the Parties (Vadrot, 2014b: 103). The multifaceted challenges of biodiversity politics and science and the difficulty of arguing the case for a new institution partly explains why the debate on a science-policy interface for biodiversity was coined as "IPCC for biodiversity": "In order to ensure coordinated actions internationally and appropriate transfer of knowledge between science and policy, it is time to establish an international or intergovernmental mechanism playing a role akin to that of the IPCC for climate change on all aspects of biodiversity" (Barbault & Le Duc, 2005: 56).

From IPCC to the Millennium Ecosystem Assessment and IMoSEB

The Intergovernmental Panel on Climate Change (IPCC) was established by the United Nations Environment Programme (UNEP) and the World Meteorological Organization (WMO) in 1988 and describes itself as "the leading international body for the assessment of climate change [established] to provide the world with a clear scientific view on the current state of knowledge in climate change and its potential environmental and socio-economic impacts".[8] The success of IPCC in popularizing climate change as an environmental problem but also as a policy issue involving policy makers and the wider public has been a major push for the debate on establishing a science-policy interface for biodiversity, particularly in the scientific community.[9] The IPCC provided not only a model for how to interface science and policy at a global level and in a complex and multilateral institutional landscape, its focus on policy-relevant global assessments did also coincide with parallel trends in biodiversity research most notably since the publication of the Global Biodiversity Assessment in 1995: the simultaneous globalization and fragmentation of environmental sciences.

Both are visible in the Millennium Ecosystem Assessment (MA, 2005),[10] which was launched by UN Secretary-General Kofi Annan in June 2001. The fact that GBA did not receive the policy-maker attention expected by the MA authors and initiators partly influenced the shift towards policy-relevance in the Assessment. MA was not conceptualized as an official intergovernmental process and the "buy-in" for the Assessment was obtained through relevant multilateral environmental agreements (MEAs) that in turn requested and supported the process. Consequently, MA was primarily aimed at meeting the needs for assessments of CBD and other biodiversity-related MEAs such as the UN Convention to Combat Desertification (UNCCD), the Convention on Migratory Species (CMS) or the Ramsar Convention on Wetlands. MA ran for a four-year period from 2001 and involved more than 1,400 scientists (Wells, Grossman & Navajas, 2006: 1). Its objective was "to derive scientifically credible, policy-relevant information for decision makers" (Norgaard, 2008: 863). Its work was based on a new conceptual framework for analysing and understanding the impact of environmental change on ecosystems and human wellbeing insofar as it "viewed ecosystems through the lens of the services that they provide to society" (Carpenter et al., 2009: 1305).

> [I]t was the first time that a comprehensive assessment of biodiversity and ecosystem services really put ecosystem services on the map and then it finished in 2005 and it was always designed to be repeatable but it was never funded in such a way that it could be repeated in that same structure or model. So there was recognition that it was useful, but there was nothing going on to repeat it.
>
> (Member of the Canadian Delegation to IPBES, Interview, June 2010 in Busan/Rep. of Korea at the third IPBES meeting)

In 2005, at its final meeting, the MA Board decided to undertake an evaluation of the utility of MA in order to determine whether it was being used by policy makers.[11] The Impact Survey conducted by Walter Reid et al. in March 2006 "provides widespread evidence that the assessment is having an impact on the intended audiences but the extent of that impact is very mixed, with some institutions, regions, countries and sectors significantly influenced by the MA while others have not been influenced at all" (Reid, 2006: 1).[12] Wells, Grossman & Navajas (2006) summarized the major achievements and weaknesses of MA, arguing that it contributed to keeping biodiversity conservation and ecosystem management on the international agenda, but that its objectives were too ambitious for a four-year project. A study conducted by the Swedish International Biodiversity Programme (SwedBio) concludes:

> [t]here was a lack of working models that could be used readily by policy makers to analyse ecosystem services, and their trade-offs with the development policies and resource allocations [...] The MA fell also short of providing convincing economic values of ecosystem services and in particular the regulating and cultural services.
>
> (UNEP/IPBES, 2010a: 4)

One important argument for a follow-up process was *inter alia* the uniqueness of the process and its impact on the formation of a genuine global community for multi-scale ecosystem assessments (Wells, Grossman & Navajas, 2006: 37). Richard B. Norgaard described this uniqueness as follows:

> In my judgment the most important product of the MA was the shared learning experience among the scientists who have participated. Because of the MA we have a cadre of scientists who have come to a deeper understanding of the complexity and sensitivity of how people and

nature interrelate and a keen sense of the direction the whole system is heading. These are the scientists who are now the most capable of informing policy.

(Norgaard, 2008: 867)[13]

One effect of the formation of this "genuine global community" was that several MA partners worked towards follow-up initiatives. In 2007, a consortium was set up to catalyse these initiatives and facilitate a coordinated MA follow-up effort, i.e. MA implementation. An MA Follow-up Global Secretariat was established and based at the UNEP Division of Environmental Policy Implementation (DEPI) with the aim of assisting in the coordination of the process, which had four main objectives: firstly, to build and improve the knowledge base on links between biodiversity, ecosystem functioning, ecosystem services and human wellbeing; secondly, to integrate the MA ecosystem service approach into decision making at all levels; thirdly, to disseminate the MA findings and its conceptual framework, tools and methodologies to relevant stakeholders; and fourthly, to explore the need, scope and modalities for a possible second global assessment (UNEP/IPBES, 2010a: 11).

MA and its follow-up were accompanied by a series of activities and initiatives building on the MA findings, such as the TEEB-study (the Economics of Ecosystems and Biodiversity), the Ecosystem Management Programme (EMP), led by UNEP and the Programme on Ecosystem Change and Society led by ICSU (UNEP/IPBES, 2010a: 11). MA and its results have influenced research agendas (Carpenter et al., 2009) and prompted the expectation "that there would then be an exploration of the possibility of creating an ongoing IPCC-like process".[14] However, that plan was put on hold because of another process initiated by the French government in 2005 as a follow-up to the Biodiversity Science and Governance Conference held in Paris. It was attended by 2,000 scientists, policy representatives and non-governmental stakeholders with the aim of identifying the mutual needs of biodiversity science and politics and to bring together the different proposals and ideas for creating an "IPCC for biodiversity". As the conference provided a platform for different interest groups that were not necessarily aware of each other, it is often seen as another important step towards the establishment of a global science-policy interface for biodiversity.[15] In his introductory speech, Jacques Chirac, French president at that time, argued that scientific consensus on the effects of environmental change on biodiversity is a precondition for developing policy-relevant knowledge. He argued that "we must deepen our knowledge on biodiversity and establish premises recognized by all scientists, so that the international community can shoulder its responsibility (Jacques Chirac, in Barbault & Le Duc, 2005: 45).[16]

Chirac's statement was inspired by the scientific community and brought forward by various circles, such as DIVERSITAS.[17] One of the main results of the conference was the initiation of the consultation process in the framework of the International Mechanism of Scientific Expertise on Biodiversity (IMoSEB), which was financed by the French government over a three-year period. In their commentary "Diversity without representation" (published in *Nature* in 2006) Michel Loreau, Alfred Oteng-Yeboah and other scientists and policy advisors (*inter alia* to national delegations to CBD COPs) argued that "[t]hese consultations should be seized as a unique opportunity to move biodiversity science and governance forward and to find new ways of resolving the crisis" (Loreau et al., 2006: 246).

IMoSEB officially started in February 2006 with the first International Steering Committee meeting and the appointment of the Executive Committee. The members of the International Steering Committee and of the Executive Committee were mostly scientists specialized in ecology, biology or botany, including both scientific researchers and experts working for governmental or non-governmental organizations. The latter were often members of the official national delegation to CBD Conference of the Parties meetings (COPs) and of other MEAs. At the first meeting, a two-step action plan was adopted and the discussion of existing mechanisms, such as IPCC, the Millennium Ecosystem Assessment (MA) and the Global Biodiversity Assessment (GBA), was aimed at looking for models to achieve a sustainable organization. From January to October 2007, six regional meetings were held in Africa, Asia, Europe, Latin America, North America and the Pacific. In order to increase the visibility of the process and link with the needs of CBD and other biodiversity-related MEAs, intermediary results were presented at different occasions, such as at the 12th CBD-SBSTTA in 2007.[18]

At the second meeting of the international steering committee in Montpellier, France, in November 2007, the final IMoSEB recommendations were adopted. The most important outcome was the recommendation that a new "intergovernmental [body] with scientific credibility, political legitimacy and relevance" was needed (IMoSEB Report, 2008: 10). The committee invited the UNEP Executive Director to collaborate with IMoSEB partners, national governments,[19] the secretariat of CBD, SBSTTA and COP bureaus in order "to convene an intergovernmental meeting with relevant governmental, and non-governmental organisations, including the relevant MEAs, academic institutions and civil society (including local communities and indigenous people) to consider establishing an efficient international science-policy interface to address the above objectives" (IMoSEB Report, 2008: 9).

From IMoSEB and the MA-follow-up to IPBES

> IMoSEB had many, many meetings to discuss that [IPCC-like structure]. They said that they didn't want this top down approach and then there was a very decisive meeting that happened in Bonn at COP 9 which I actually witnessed.
> (Global Environmental Facility [GEF], Department of Natural Resources, Head, Interview, June 2010 in Busan/Rep. of Korea at the third IPBES meeting)

Current literature on IPBES often refers to the new science-policy interface as a result of merging of IMoSEB and the follow-up of the Millennium Ecosystem Assessment (Esguerra, 2015; Granjou et al., 2013; Koetz, Farrell and Bridgewater, 2012; Neves, Turnhout and Lijster, 2014; Vadrot, 2014a; Vohland & Nadim, 2015). After the MA publication in 2005 and publication of the Fourth Global Environmental Outlook (GEO 4), the debate on establishing a framework for ensuring the regular production of global assessments of the state of biodiversity worldwide continued within different stakeholder groups and communities. However, the question remained on how and under what conditions this could be done, particularly with regard to the different approaches of IMoSEB and MA.

"Unfortunately, the initial plans for IMoSEB seemed to have few of the features that we believed made the MA most useful and relevant to decision-makers. Most importantly, the MA was framed around the question of 'how do changes to biodiversity and ecosystem services affect human well-being?' and was thus as strongly focused on development as environment. IMoSEB however has focused more narrowly on biodiversity in its own right. It was also disappointing that the IMoSEB effort was developed without any linage to the MA. Our thinking in the MA had been that after governments had experience with the MA they would be more likely to see that an ongoing process built on the MA would be acceptable. IMoSEB must instead 'prove' its utility just as the MA has done before it is likely to be accepted."[20]

The ninth Conference of the Parties of CBD (COP), which took place in Bonn, Germany, from 19–30 May 2008 provided a platform for negotiating the different initiatives and proposals. Both, the MA follow-up process and the recommendation of IMoSEB were issues of debate in the COP Working Group II of 27 May. The respective text was proposed by the CBD Secretariat and the SBSTTA recommendation XII was to undertake ecosystem assessments every 5–10 years (MA follow-up) and to include a new text referring to IMoSEB (UNEP/CBD/COP/9/13).

The need for the establishment of a new body was highly contested, particularly by Brazil, Argentina and the African Group that opposed any reference to IMoSEB (ENB, 2008b: 1). An interview with a Brazilian delegate revealed the reason for Brazil's strong opposition to a new intergovernmental body. It was based primarily on the belief that strengthening existing mechanisms would make more sense than creating a new body.[21] Smaller developing countries argued that, as enough scientific knowledge was available, any additional institution would not contribute to strengthening the role of traditional and indigenous knowledge and was hence not needed (Vadrot, Heumesser & Ritzberger, 2010: 205).

Contrary to this position, Canada, EU, Mexico, Peru and Malaysia were in favour of a new body, welcoming the IMoSEB initiative and merging of the MA follow-up process and IMoSEB. A consensus between the different positions was reached outside the plenary. "It was actually a negotiation between certain key countries in this process, Brazil included. Brazil, Mexico, France, USA, and some representatives from Africa. In that meeting, finally the IMoSEB people said "OK. We are ready to close down this process" (IUCN – International Union for the Conservation of Nature, Head of Ecosystem Management Programme, Interview, June 2010 in Busan/Rep. of Korea at the third IPBES meeting). The IUCN and representatives of these countries explained the change of their positions as follows: "We are ready to consider something along this line if it is constructed bottom up, and if it is to be something that is not driven by the perspective of developed countries" (GEF). It was UNEP that proposed to combine the expressed need for regular global assessments with IMoSEB and "to develop an intergovernmental and multi-stakeholder approach to strengthen the science-policy interface on biodiversity and ecosystem services" (ENB, 2008b: 2).

COP Decision IX/15²² (2008) was the beginning of a five-year negotiation process under the auspices of UNEP and involving stakeholders from more than 100 countries, scientific communities and NGOs, with the aim of deciding whether and, if yes, how the science-policy interface for biodiversity should be strengthened.

Negotiating the establishment of IPBES

The negotiations leading to the decision that a new body should be established to strengthen the interface between science and policy for biodiversity included three Intergovernmental and Multi-Stakeholder Meetings that took place in 2008, 2009 and 2010, and two plenary sessions in 2011 and 2012. Whilst the first three meetings were focused on the question of whether and how the science-policy interface should be strengthened, the two plenary meetings addressed the modalities and institutionalization of a new intergovernmental platform. In the following, the negotiation dynamics will be described as shown in Figure 3.1.

Towards the Busan Outcome – whether to establish a new body

From 10 to 12 November 2008, an *Ad hoc Intergovernmental and Multi-Stakeholder Meeting on an Intergovernmental Science-Policy Platform on Biodiversity and Ecosystem Services* (IPBES) took place in the International Convention Center in Putrajaya, Malaysia. According to the German Federal Environment Minister and COP 9 President Sigmar Gabriel, the meeting gave decision makers "the first opportunity after COP 9 to prove that we are serious about our decisions taken in Bonn. We now can take a big step forward and demonstrate our commitment to enhancing the quantity and quality of scientific information available to institutions and organisations at different levels." In his message to participants published one week before the Putrajaya meeting, he

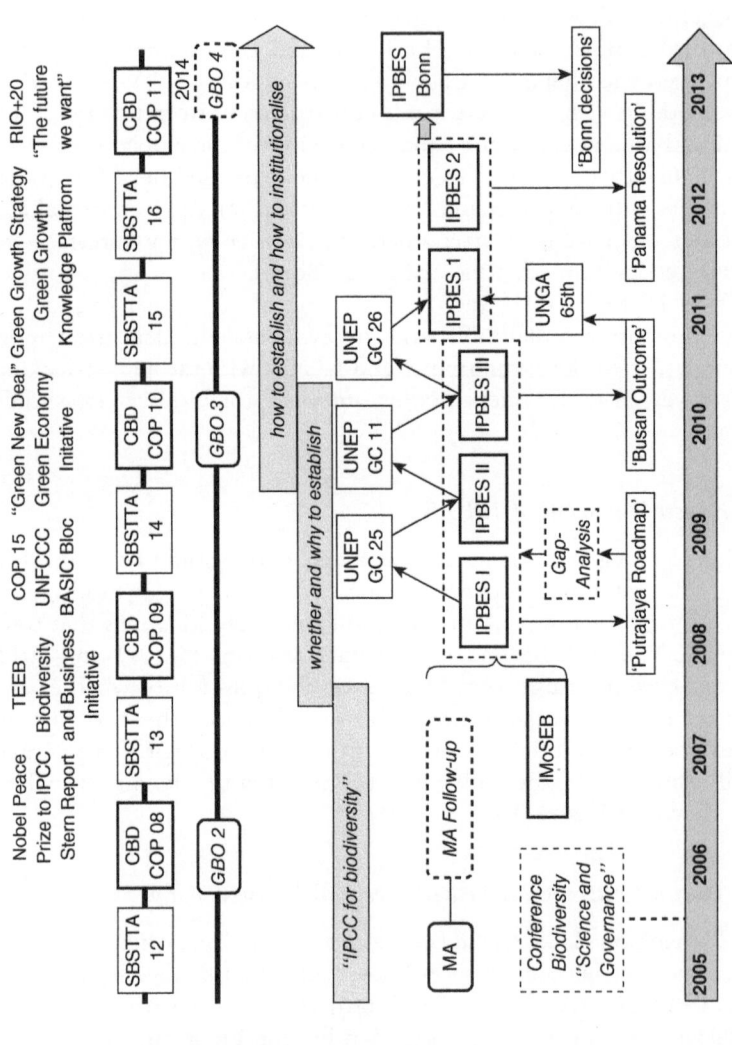

Figure 3.1 Historical development of IPBES negotiations
Vadrot (2014b: 139)

emphasized the need for "strong recommendations to be addressed at the UNEP Governing Council in February 2009 in order to establish the IPBES in the same year".[23]

The meeting was attended by 176 participants from 78 countries and 25, mostly scientific, organizations (ENB, 2008a: 1).[24] Some of the participants were previously involved in the IMoSEB and MA follow-up, but other delegates did not have any knowledge about the previous processes and initiatives.[25] As opening speaker, the UNEP Secretary General Achim Steiner emphasized the importance of the meeting, claiming "that the current meeting represented an endeavour to understand how science, research and knowledge could best be brought to the policy level with regard to the future use and management of biodiversity and ecosystem services" (UNEP, 2008: 1). In his speech, he referred to the economic impact of biodiversity loss and ecosystem degradation and the possibility of reconciling economic development with conservation, acknowledging however that "the rationalization of biodiversity in economic terms remained [...] an uncomfortable notion for many" (UNEP, 2008: 1).

Despite the opposition of Brazil, the principles of the UNEP Governing Council were adopted for the meeting in order to ensure its formal character and general acceptance of the results. Brazil was against the application of the consensus principle as it was opting for a rather informal consultative meeting. Since the beginning the conflicts between the different parties at the COP 9 negotiations exacerbated the debate and the different expectations from a new body and its main tasks came to the fore. Whereas the USA, EU and Switzerland focused on the quality of the end products of a new body in terms of scientific credibility and saliency (i.e. regular global and sub-global assessments), Brazil, in the first place, argued that a new platform should enable developing countries to build their own scientific capacities and recognize the value of traditional knowledge (ENB, 2008b: 6). Hence, Brazil opted for a framework suitable to support the implementation of CBD's triple objective, whereas the USA opted for a new body that should focus merely on biodiversity conservation and thus not be established under CBD or CBD-SBSTTA (ENB, 2008a: 7).

In light of the proposition to conduct a gap analysis on the actual need for a new institution,[26] delegates "agreed that no recommendations as such would be adopted, but that the Chair's summary [...] would serve as the outcome from the meeting" (UNEP/IPBES, 2008: 8). The Chair's summary was published as an Annex to the report ("Putrayaja Road Map"). The Annex was criticized particularly by Brazil for not reflecting the views of the meeting but only those of the Vice-Chair, Robert Watson (ENB, 2008a: 8).[27] The summary provided by the Chair read that

> mechanisms to improve the science-policy interface for biodiversity and ecosystem services for human well-being and sustainable development should continue to be explored [...] and that a gap analysis should be

undertaken for the purpose of strengthening the science-policy interface and that a preliminary report should be made available at the twenty-fifth session of the Governing Council/Global Ministerial Environment Forum.

(UNEP/IPBES 2008: 12)

Four months later, the UNEP Executive Director reported on the outcome of the meeting to the 25th session of the UN Governing Council (UNGC) held in Nairobi from 16 to 20 February 2009. He presented preliminary results of the gap analysis. Based on the analysis, he proposed that UNEP convene a second intergovernmental multi-stakeholder meeting on the creation of IPBES in the light of the full results of the gap analysis (UNEP/IPBES, 2008: 12).

The second *ad hoc* intergovernmental and multi-stakeholder meeting on setting up IPBES was held in Nairobi from 6 to 9 October 2009.[28] Prior to the meeting, Harold Mooney and Georgina Mace, both scientific advisors to IMoSEB and respectively Chair and Vice-Chair of the Science Committee of DIVERSITAS, had published an editorial in *Science* arguing that

[T]he meeting is a chance to boost international action, based on strong scientific evidence [...] Unlike the UN Framework Convention on Climate Change, which has the Intergovernmental Panel on Climate Change, these environmental conventions lack a pre-convention science assessment and have no provision for subsequent government-endorsed, independent science. The meeting in Nairobi will debate, among other issues, how best to make up for this crucial omission.

(Mooney & Mace 2009: 1474)[29]

The second IPBES meeting started with the presentation of the key findings of the gap analysis, which served as a baseline for the subsequent debate. During the debate, it became clear that capacity building should be seen as one pillar of a future IPBES, particularly because of its important role for developing countries, based on the idea that capacity building is a means by which research infrastructures could be strengthened in their countries (see chapter 10). Despite the strong support for a new intergovernmental mechanism, there was no decision on its establishment as a result of the deliberations. Like at the end of the first meeting, the Chair provided a summary of the debate that would serve as baseline for a third and arbitrative meeting on IPBES. According to one interviewee, "in Nairobi most parties said that they would be supportive of this process, but that they would not at that meeting be able or have the mandate to get their governments to negotiate" (Interview, June 2010 in Busan/Rep. of Korea at the third IPBES meeting, Global Environmental Facility (GEF), Department of Natural Resources, Head). From the point of view of another interviewee, the meeting resembled any negotiation and any new mechanism. People would naturally all be suspicious. You have to check if we don't already have enough and if we are not just creating another burden. These things always come with money, so where is the new money going to come from? [...] I remember that the second meeting was a very difficult meeting. Difficult in the way of that there was a lot of suspicion – on what this 'animal' is all about.

(South Africa Delegation, Member, Interview, June 2010 in
Busan/Rep. of Korea at the third IPBES meeting)

The results of the second IPBES meeting were presented at the 11th special session of the UNEP Governing Council, which took place from 24 to 26 February 2010 in Bali (Indonesia). This time, the UNEP Executive Director was mandated to organize a final meeting on the establishment of IPBES and to "undertake, within available resources, action for the implementation of outcomes of the third meeting and relevant resolutions that might be adopted by the General Assembly on this matter, and requested to report thereon to the Governing Council at its 26th session" (UNEP/GCSS, 2009: 4).

The third meeting, which led to the so called "Busan Outcome", was held from 7 to 11 June 2010, in Busan (Republic of Korea) to acknowledge the International Year of Biodiversity. Twice as many people attended the meeting as compared to the first meeting. Some 232 delegates representing 85 countries, and 21 NGOs/IGOs negotiated for five days to facilitate the establishment of IPBES.[30] In their comment previously published in *Science*, Anne Larigauderie, Director of DIVERSITAS, and Harold Mooney, a renowned American ecologist, Co-Chair of MA and Chair of the Science Committee of DIVERSITAS, appealed to the scientific community, arguing that "scientists can no longer hope that their work will somehow be used by policy makers. They should try to understand how policy works at local, national or international levels, dialog with users of information to develop together an understanding of knowledge needs, and try to adapt and focus their work to these needs" (Larigauderie & Mooney 2010: 14). In their comment entitled "Wanted: An IPCC for Biodiversity", the editors of *Nature* heralded the negotiations: "Moves are now afoot to establish a body to review the science and anticipated effects of changes in biodiversity, reminiscent of the Inter-governmental Panel on Climate Change (IPCC). Next week in Busan, South Korea, representatives from governments around the world will decide whether to create such a panel" (*Nature*, 2010: 525).

The third meeting differed from the first two as it was designated as the "final" meeting. Compared to the previous meetings, it was much better prepared and required, for the first time, the credentials of government representatives.[31] In turn, NGO and IGO representatives lost their equal footing with government representatives, so they were no longer part of the consensus-making actors.[32]

After the keynote speech, again delivered by Achim Steiner, the deliberations during the next two days were developing in a quite enthusiastic spirit.

Thereafter, however, the negotiations became increasingly complicated. The difficulties started with a proposition put forward by the Chair, Mr Kim Chan-woo (Republic of Korea), Asian and Pacific group. He proposed to reconvene the drafting group on the "Busan Outcome" document. USA, EU and China held the opinion that this was far too early as there was "no basis for a decision yet" (ENB, 2010a: 2). As in the first meeting, the Vice-Chair Robert Watson was criticized by many parties for being too influential on the negotiations and the South Korean Chair for acting against UN rules and procedures, with the reason being that the Chair proposed to reconvene the drafting group on the "Busan Outcome" document, which for many countries – USA, EU and China at the forefront – did not reflect the state of the negotiations, or, in other words: "there is no basis for a decision yet" (ENB, 2010a: 2). The USA continually criticized the Chair for not being in line with UN rules of procedure. The EU turned out to be in a difficult position, being an IPBES supporter, but reluctant with respect to the strong emphasis on capacity building by countries such as Brazil and most African states. Furthermore, the conflict lines between developed and developing countries intensified. Whilst most developing countries stressed the importance of regular financial contributions to IPBES, EU and USA argued in favour of voluntary contributions. The EU proposed that the funding should also include other sources "such as the private sector and foundations". As a reaction, Brazil reproached the EU for not shouldering its responsibility. At this stage, a clear line of demarcation within the plenary was discernible and the disappointment of the scientific community was visible – the legal status and the means by which to establish a new body was still unclear.

The relationship between IPBES and the existing bodies like CBD, CBD-SBSTTA and other entities remained unclear. Another controversial issue was determining who should be entitled to propose research topics and knowledge brokerage services to IPBES. Furthermore, "new emergent issues" remained a critical topic: Should IPBES identify new emerging issues and new research areas, or does this go beyond the IPBES mandate? And, finally: Should IPBES become policy prescriptive or restricted to policy relevance?

Other issues put a strain on the debate as well. The document basically was mostly about the intergovernmental actors. So what role is left for "other actors", such as scientific networks, business and industry? Furthermore, what importance is assigned to the significance of other non-scientific knowledge sources, the role of traditional knowledge and the contribution of local and indigenous communities? Related to this: How can recognition of the regional character of biodiversity and, hence, of regional differences be ensured? Another topic relevant for developing countries was the ownership of knowledge, i.e. for these countries, free and open access to data and information and knowledge transfer was essential.

China, Brazil and other delegations, including Colombia and Iran, made it clear that they did not wish IPBES to identify research needs or gaps and that the identification of new emerging issues (such as biofuels and bioengineering) should not be discussed outside CBD COPs and SBSTTA. Most delegates "were still concerned that the identification of 'emerging' issues would invite political subjectivity, but that, as noted by several delegates, removing the term would limit IPBES' ability to inform policy makers of new trends" (ENB, 2010c: 5). Even though the identification of new emerging issues was previously – *inter alia* in the IMoSEB recommendations and the first two IPBES meetings – viewed as an important pillar of a new body interfacing science and policy for biodiversity, the Parties agreed that no reference should be made to this in the Busan Outcome.

In reaction to this, the ICSU representative, Charles Perrings, claimed that no rules and regulations should be drawn up on this topic as the identification of new emerging issues would be critical for the work of the IPBES. He pointed to the vital role of screening of biodiversity and ecosystem services. In order to stress the need for scientific credibility and independence, Perrings appealed to "the scientists among delegations" to focus on the basic functions of a future body because the planet was at risk.

Especially towards the very end of the negotiations, debates between the Chair and the delegations became quite agitated.[33] However, the Secretariat presented the draft of the Busan Outcome (ENB, 2010b: 8). The participants finally concluded "having now reached agreement [...] that an intergovernmental science-policy platform for biodiversity and ecosystem services should be established to strengthen the science-policy interface for biodiversity and ecosystem services for the conservation and sustainable use of biodiversity, long-term human well-being and sustainable development" (UNEP/IPBES, 2009: 5). The work programme consisted of four pillars: knowledge generation, assessment, policy support, capacity building In Paragraph 7 the *modus operandi* of IPBES and the principles guiding the work of the platform were further specified.[34]

The Busan Outcome was presented at the 65th session the UN General Assembly. UNEP got the mandate to take appropriate action to establish the platform, for which two plenary meetings were organized.

How to establish and institutionalize?

The first two IPBES plenary meetings were designated to determine the modalities and institutional arrangements for an intergovernmental science-policy platform on biodiversity and ecosystem services, including *inter alia* its legal basis, the physical location of its secretariat and its administration.

The first session was held from 3 to 7 October 2011, at the UNEP headquarters in Nairobi (Kenya). Robert Watson, re-elected as Chair of the meeting, stated that the IPBES modalities should be "put in place as a matter of urgency while *getting them correct*" (ENB, 2011: 2). In his opening speech, Achim Steiner pointed out that "there is a massive scientific community looking at you and expecting that the IPBES will finally be established". Previously, Charles Perrings and other scientists associated with IUCN, ICSU and DIVERSITAS published a commentary in *Science* (Perrings et al.,

2011), which was followed by two comments from other scholars, criticizing the strong focus on global scientific assessments set out in the Perrings text (Briggs & Knight, 2011; Hulme et al., 2011). The conflicting views between those groups were related to the role and design of scientific assessments and the types of knowledge that should be recognized. Whilst Perrings et al. (2011) argued in favour of an IPCC for biodiversity with a strong emphasis on the assessment of ecosystem services and their value, Hulme et al. (2011) and Briggs and Knight (2011) pointed to the need for decentralized knowledge-production, for regional and local assessments and the recognition of non-scientific knowledge. Similar discrepancies in the understanding of what knowledge IPBES should generate were visible during the first session negotiations. But, before these turned out to challenge the negotiations, another fundamental question arose: Did the Busan Outcome and the statement by UNGA establish IPBES or not?

The decisions taken regarding the status of IPBES at different international meetings and conferences, such as within the framework of CBD COP 10,[35] the 185th session of UNESCO,[36] the 65th session of the United Nations General Assembly,[37] FAO[38] and CITES,[39] reveal uncertainty and reluctance among parties regarding the role and status of IPBES in the global biodiversity governance system in general, and the individual conferences, programmes and conventions in particular.[40]

In order to clarify the situation, UNEP sent a request to the Office of Legal Affairs of the United Nations (OLA). At this stage, some delegations started questioning the Busan Outcome. The USA, for example, had concerns regarding the role of IPBES among parties. Ghana wanted the status of capacity building to be clarified. Bolivia "reserved the right to re-open discussion on particular items" and Colombia "expressed concerns on how to operationalize the IPBES functions" (ENB, 2011: 2). The response of OLA arrived at the third negotiation day. It confirmed that the UN General Assembly resolution had not established IPBES.[41] Subsequent negotiations on the work programme showed that seemingly insurmountable differences in the understanding of the kind of knowledge that IPBES should generate or produce would further delay the IPBES establishment.

Switzerland and Indonesia pointed to the importance of assessing ecosystem services and their benefits within the IPBES work programme. Norway, Peru and Fiji underlined the importance of recognizing local and traditional knowledge and the Plurinational State of Bolivia expressed its worries regarding the emphasis on the economic value and ecosystem services (see Brand & Vadrot 2013; Vadrot, 2014b). Brazil argued that, "We need a broad range of understandings on biodiversity. If IPBES will help to provide mechanism to show the value of biodiversity this needs to be done. At the moment the value is 0. We need to change this."[42] In turn, Bolivia emphasized that, "The issue of economic valuation was an issue why we did not want to establish the platform. We have defended this in other forums. We are very happy with the fact that nature has no value, that we can benefit from nature and breathe the air without paying. The value you will put on this will never be appropriate" (statement by the Bolivian government representative).[43] The concerns raised by Bolivia proved to further challenge the establishment of IPBES.

When the final report on the negotiations was presented by the Chair, Bolivia refused to agree to the text, because it did not include a reference to the concerns on the ecosystem service concept raised by Bolivia. The Bolivian delegate formulated a passage that was included in the final report and endorsed it. In the end, however, no agreement could be found on the resolutions presented by the Chair. Against this background, the parties agreed not to formulate a resolution on the establishment of IPBES at all and to postpone any decision to the second session that was to be held in Panama City from 16 to 21 April 2012.

After this heated first plenary meeting, most representatives expected the second meeting to result at least in the choice of a physical location for IPBES and an agreement on some key aspects of the modalities and institutional arrangements for the new platform. Proposals for hosting IPBES were submitted by the governments of France, Germany, India, Kenya and the Republic of Korea in December 2011. The individual proposals were presented at the third day of negotiations by government representatives of the submitting states (UNEP/IPBES 2012a).

Germany made the most generous offer, promising that US$1.3 million would be paid to the IPBES trust fund annually, US$850,000 provided for financing conferences, travel costs and studies, and US$6.5 million made available annually for capacity-building activities. Germany won the vote on the fourth day of negotiations. The voting was organized in four rounds. There was no clear majority after the first round and India, the candidate with the fewest votes (five or six), was removed from the ballot. The first round showed that the Republic of Korea received the most votes (30), followed by Germany and Kenya with 20 votes each, and France with only around seven votes. Consequently, France was excluded in the second round. In the third round, Kenya was removed as it received the fewest votes. Finally, Germany received slightly more votes than the Republic of Korea in the third round (German Delegation, Member, telephone interview, June 2012).

Regarding the host institutions, UNEP, UNDP, UNESCO and FAO provided the participants with a joint proposal on how to administer IPBES. However, no agreement could be found as many governments claimed not to have enough information on how the work would be organized. Furthermore, some countries argued that only one organization should be the host institution, with Kenya and other African countries arguing that IPBES would fall under the authority of UNEP. Subsequent debates on the subsidiary bodies and the composition of the Bureau also failed to result in joint proposals even though the meeting continued late into the night (ENB, 2012a:2).

On the last day of negotiations, a proposal for the interim and intercessional work of the "multi-disciplinary expert panel" (MEP)[44] was developed and basic points such as regional representation, the composition of the

Bureau and the function of the Chairs and Vice-Chairs were tackled. In view of a lack of agreement on capacity building, assessments and other IPBES deliverables, it was finally agreed that "the intercessional process would be undertaken with broad participation from the scientific and policy community and knowledge holders; emphasize balanced representation of developed and developing countries and economies in transition; and elaborate on how the MEP would be permanently structured" (ENB, 2012b:1).

On Saturday 21 April 2012, the participants came together to agree on the text of the resolution. Again, diverging views hampered the adoption of the proposed text (UNEP/IPBES.MI/2/7). There were many controversial issues, e.g. UNEP's role as sole host organization and potential transformation of IPBES into an entity independent from the United Nations through the UNEP Governing Council. This was particularly contested by the African Group, Bolivia, and Indonesia. In this regard, other parties referred to IPCC as a role model. Another contentious issue was the nature of IPBES decisions. Even though there was agreement that these should be "non-legally binding" their domain was still unclear.[45] However, in the end the resolution was adopted and IPBES was established with the Panama Resolution (UNEP/IPBES 2012b, see Box 3.1).

Box 3.1 Intergovernmental Science-Policy Platform on Biodiversity and Ecosystem Services

We, the representatives of the Governments listed hereunder, assembled on the occasion of the second session of the plenary meeting to determine modalities and institutional arrangements for an intergovernmental science-policy platform on biodiversity and ecosystem services, convened in Panama City, Panama, from 16 to 21 April 2012,

Recalling the Busan outcome, adopted at the third *ad hoc* intergovernmental and multi-stakeholder meeting on an intergovernmental science-policy platform on biodiversity and ecosystem services, convened in Busan, Republic of Korea, from 7 to 11 June 2010,

Recalling also General Assembly resolution 65/162 of 20 December 2010 and decision 26/4 of 24 February 2011 of the Governing Council of the United Nations Environment Programme,

Recalling further the work of the first session of the plenary meeting to determine modalities and institutional arrangements for an intergovernmental science-policy platform on biodiversity and ecosystem services, convened in Nairobi from 3 to 7 October 2011,

Recalling the Rio Declaration on Environment and Development,

1   Decide to establish an independent intergovernmental body to be known as the Intergovernmental Science-Policy Platform on Biodiversity and Ecosystem Services;

2 Also decide that the first session of the Plenary of the Platform will decide on the link between the Platform and the United Nations system;

3 Further decide, in order to fully operationalize the Platform:

(a) That the functions, operating principles and institutional arrangements of the Platform are to be those set out in appendix I to the present resolution;

(b) To request the secretariat of the United Nations Environment Programme to facilitate the Platform until the secretariat of the Platform is established, with a view to its being administered by one or more of the following: the United Nations Environment Programme, the United Nations Educational, Scientific and Cultural Organization, the Food and Agriculture Organization of the United Nations and the United Nations Development Programme;

(c) That the seat of the secretariat of the Platform is to be located in Bonn, Germany;

(d) That the rules of procedure, as contained in appendix II to the present resolution, are to be used by the Platform and may be modified by the Plenary of the Platform;

(e) That the present resolution and any future decisions of the Platform have a legally non-binding nature.

Concluding analysis: policy change and conflicts in the institutionalization of IPBES

The aim of this chapter was to provide an overview of the history of the establishment of IPBES from the early beginning to its establishment in Panama in 2012. As shown in Figure 3.1, I distinguish between three main stages of the IPBES institutionalization. Firstly, the early beginnings, where the rationale for establishing an "IPCC for biodiversity", i.e. a global and institutionalized science-policy interface addressing biodiversity and related issues in a systematic way, was formed by the intervention of various scientific communities and complemented by the increased perception of governance failure in the framework of CBD and the overall landscape of multilateral environmental agreements (see chapter 2). This first stage started with the recognition of biodiversity loss and its sustainable use as a policy issue and as a scientific object in the 1990s and peaked when the IMoSEB process was launched in 2005 and merged with the MA follow-up in 2008. The second stage covers the three multi-stakeholder meetings characterized by debates around the question on "whether and why to establish" a new institution. In this phase, conflicts that arise in international biodiversity politics in general and CBD in particular overlapped with the objective shared by most states to establish an "IPCC for biodiversity", most notably visible in conflicts around capacity building, technology transfer[46] and the role of traditional knowledge.

Thirdly, the stage was reached where the question on "how to establish and how to institutionalise" dominated the overall debate and triggered conflicts related to the rules and procedures of multilateral environmental politics mirroring power constellations in place since the establishment of CBD. Even though these questions had been present since the beginning of the IPBES negotiations, the pending legal status of IPBES identified at the first plenary session in Nairobi in 2011 represents the starting point of this third stage, where the IPBES functions and work programme was still being debated, but the need for a new body was no longer questioned. This third stage did not end with the establishment of IPBES in Panama in 2012, insofar as the institutions, instead of being reified, were constantly changing assemblages of strategies, interests, values, routines and expectations of the various actors involved. The described stages overlapped and discursive elements of each stage were present at every other stage. A full understanding of the establishment of IPBES thus requires an analytical framework going beyond a positivist understanding of institutional change as a linear process.

In my introduction, I started from the assumption that the creation of an "IPCC for biodiversity" has to be understood in relation to the challenges of coping with policy dynamics, particularly if the ontology underlying the analysis is based upon an understanding of agency and structure as dialectically co-evolving and mutually selectively structured. This means that the emergence of an institution and the various policy dynamics in place have arisen out of and have an impact on the formation of the actors' interests and "on the way they act in relation to their interests" and expectations about the future (Agrawal, 2005: 98f.). This last conclusion aims at systematically grasping the interdependencies between institutional and epistemic change fostering and characterizing the establishment of IPBES in terms of narratives.

Three types of narratives and the spatiotemporal pattern[47]

An extensive analysis of policy documents, negotiation reports and interview material reveals that the early debates on whether and how to establish a new body were structured around three types of narratives regarding the *raison d'être* of IPBES and its role within the broader institutional landscape of international biodiversity politics (Vadrot, 2014a).

Narratives are enduring and reveal the particularities of the science-policy interface, along with the role that science, scientific communities and individual scientists played throughout the whole process. They help us discover which kinds of rationalities were used to make the IPBES happen, to identify both the explicit and implicit conflicts and relate them to the 'bigger picture', i.e. the epistemic selectivities of international biodiversity politics and the 'global paradigm of valorisation'.

(Vadrot, 2014b: 9f.)

The narratives identified are both formed by and constitutive of the expectations of different actors, institutions and individuals directly and indirectly participating in the IPBES negotiations. They allow for decryption of particular types of interactions between actors and across different policy-making levels triggered by the potentiality of a new institution characterized: firstly by the controversies between individual nation states and the UN bodies around juridical and governance issues; secondly by the controversies anticipated by different scientific communities; and thirdly by the controversies between different knowledge holders, particularly around the notion of valuation and the ecosystem service concept.

The first type of narrative ("Better Politics" and "Better Policies") relates to international biodiversity politics and the implementation deficit of CBD. It is about the role a new science-policy interfacing body could play in fostering the implementation of CBD and the expectation that a common knowledge base could contribute to solving conflicts deriving from diverging problem understandings in relation to biodiversity (loss). Throughout the multi-stakeholder meetings, the expectation that IPBES could serve as a potential catalyst to create synergies among the six biodiversity-related conventions and the policy issues they address became visible as well. However, the relationship between CBD and other organizations, namely the World Trade Organization (WTO) and the World Intellectual Property Organization (WIPO), was not at all addressed and hence the potential role of IPBES in this regard was neglected.

The second type of narrative ("Scientific evidence and usable knowledge" and "How to convince policy makers to act") is about the interface between science and policy and the perceptions and expectations of policy makers towards scientists and scientific knowledge on one hand and of the scientific communities towards policy makers and governance processes on the other. A closer look at these narratives reveals the epistemic selectivities in place, namely the particular and specific role of science in commodifying biodiversity and the prevailing dominance of Western ideals of scientific knowledge production conflicting with traditional views on nature.

The third type of narrative ("Mainstreaming Biodiversity" and "Valuing Biodiversity") reveals expectations regarding the impact of IPBES on biodiversity knowledge (production and use) and the assumption that more coherent, interdisciplinary, policy-relevant and holistic approaches to address biodiversity could contribute to halting biodiversity loss by providing additional arguments for why to protect nature. In addition, the expectation that the (monetary) value of biodiversity and ecosystem services could contribute to enhancing the effectiveness and efficiency of related policies and politics became increasingly visible in the IPBES process. In this regard, the problem of scientifically addressing the biodiversity loss issue preempts the assumption that the focus on ecosystem services and on valuing biodiversity would increase the likeliness that policy makers and the broader public will become aware of the importance of biodiversity conservation.

The identified narratives mirror the particular interrelation between the epistemic and institutional dimensions of interfacing conservation science and biodiversity policy. They also interrelate with developments in international biodiversity politics and policies constituting the spatiotemporal patterns within which IPBES emerged. International biodiversity politics and policies are characterized by a fundamental conflict between the further commodification of nature, which is being surveyed, catalogued and is the subject of patenting attempts (in particular by the pharmaceutical industry), and a less materialistic approach that draws on wider – not only scientific – knowledge. However, current environmental politics and science increasingly base their approaches on the premises of ecological economics, providing a rationale for reconciling the seemingly opposing objectives of nature conservation, poverty eradication and economic growth. The so-called "Green Economy" concept not only appeared as contact point with the negotiations of the United Nations Conference on Sustainable Development in Rio de Janeiro (June 2012), but also mirrored and prearranged the epistemic and institutional framework of the IPBES establishment strengthening approaches provided by scientifically framed policies (e.g. the increasing reference to the ecosystem service concept) against the background of a perceived dysfunctioning biodiversity governance system. The narratives mirror the streamlining of the way in which biodiversity politics can be discussed by effectively coupling governance failure and ineffectiveness with the need for common knowledge (methodologies, concepts and terms) and the objectives of credibility, salience and legitimacy, i.e. increasingly in terms of how to value biodiversity and ecosystem services.

Notes

1 CBD was opened for signature in 1992 at the United Nations Conference on Environment and Development (UNCED) and entered into force in 1993. Two years later, the Parties to the CBD established the Subsidiary Body for Scientific Technical and Technological Advice (SBSTTA) to the Convention on Biological Diversity (CBD) under Article 25 in 1992. Its establishment was contested because of the concern of some Parties that SBSTTA could exert a dominating influence on negotiations within the scope of CBD (Koetz et al., 2008: 511).

2 The heading is derived from the publication by Loreau et al. (2006), where a group of natural scientists and policy advisors argued the case for a global science-policy interface for biodiversity.

3 The Forum took place under the auspices of the American National Academy of Science and the Smithsonian Institute.

4 According to Di Castri and Younès, two of them were "of more official nature": The more extended one is that of the United Nations included in the Convention on Biological Diversity: "The variability among living organisms from all sources

including, inter alia, terrestrial, marine and other aquatic ecosystems and the ecological complexes of which they are part; this includes diversity within species, between species and of ecosystems." The shortest definition of all is that of the Global Biodiversity Strategy (World Resources Institute, IUCN and UNEP, 1992) which regards biodiversity as "The totality of genes, species, and ecosystems in a region" (Di Castri & Younès 1996: 1). The authors noted that both of them refer to the three main components "genes, species and ecosystems".

5 The International Council for Science
6 International Union for the Conservation of Nature [IUCN], Head of Ecosystem Management Programme, Interview, June 2010 in Busan/Rep. of Korea at the third IPBES meeting.
7 Empirical analyses of the use of the term in scientific literature seems to support the critical comments on the fuzziness of the term. "In 1988, biodiversity did not appear as a keyword in Biological Abstracts, and biological diversity appeared once. In 1993, biodiversity appeared 72 times, and biological diversity 19 times" and so on (Takacs 1996: 39). Once the terminology is set, it is hard to exclude anything from it and subsequently "biodiversity" would include all biological entities (e.g. Norton, 2001).
8 www.ipcc.ch/organization/organization.shtml
9 "For a long time there has been the IPCC and everybody was looking at that as there was no similar body for biodiversity and ecosystem services" (International Union for the Conservation of Nature [IUCN], Head of the Ecosystem Management Programme, Interview, June 2010 in Busan/Rep. of Korea at the third IPBES meeting).

> I think if you go way back to where it started; it started where people had IPCC-envy. They took a look at IPCC and they saw what a fantastic job they had done from moving the debate. These people got a Nobel-prize. All of a sudden people literally said: Hey we need one in biodiversity. So it started off as one group of thought inside of the scientific community.
> (Canadian Delegation to IPBES, Head, Interview, June 2010 in Busan/Rep. of Korea at the third IPBES meeting)

> I think there are a lot of assumptions riding on this premise, that if we have something much like IPCC for biodiversity that policy makers will pay attention and therefore they are going to start mainstream biodiversity and ecosystem services and insert this into their national policy making.
> (Global Environmental Facility [GEF], Department of Natural Resources, Head, Interview, June 2010 in Busan/Rep. of Korea at the third IPBES meeting)

10 The relationship between IPCC and a science-policy interface for biodiversity was not only part of the debate on how IPBES should be designed or not. The reference to IPCC was an argument for why to institutionalize the science-policy interface for biodiversity used by representatives of the scientific communities in several comments written for Nature, Science and other scientific journals (Loreau et al., 2006; Perrings, 2010; Larigauderie & Mooney, 2010). Most recently, Thomas M. Brooks et al. published a paper pointing to the necessity of clearly differing between IPCC and IPBES (Brooks, Lamoreux & Soberón, 2014). For another line of argument focussed on the necessity for a reflexive turn in the governance of global environmental expertise, see Beck et al. (2014).
11 www.publications.parliament.uk/pa/cm200607/cmselect/cmenvaud/77/77we05.htm
12 See also the first report of the UK Select Committee on Environmental Audit www.publications.parliament.uk/pa/cm200607/cmselect/cmenvaud/77/7707.htm#note15

13 Norgaard even argues that MA "serves as an existence proof, albeit a highly select one, for the possibilities of deliberative democracy for reaching shared understanding on a larger scale, for developing an informed electorate capable of providing the political will to sustain humanity and life" (Norgaard, 2008: 868).

14 www.publications.parliament.uk/pa/cm200607/cmselect/cmenvaud/77/77we05.htm

15

> I think it was an idea coming from the scientific community, not from France, from other countries and before this initiative all the countries, like Norway, tried to initiate an intergovernmental panel. I don't remember the name, but it was like the intergovernmental panel on climate change and it was killed by different countries. So it was an idea floating more or less in the community and we organised this important meeting in France "Biodiversity science and governance" where we joined different types of opportunity to propose this kind of thing.
>
> (French Delegation, Director of IMoSEB, Interview, May 2008, in Bonn/Germany at CBD COP 9)

16

> That was a political statement, which was prepared within the scientific community. When the president says something, this "something" comes to him from various circles and in France this was the scientific community. In France at the time they really pushed for that. Once it was in the political realm and declared at the international science policy conference, then it really created enough of a momentum to give the scientist a mandate to start organizing the IMoSEB process.
>
> (*DIVERSITAS*, Director, Interview, June 2010, in Busan/Rep. of Korea at the third IPBES meeting)

17

> I think there are a lot of assumptions riding on this premise that if we have something much like IPCC for biodiversity that policy makers will pay attention and therefore they are going to start mainstream biodiversity and ecosystem services and insert this into their national policy making. [...] This grew out mostly as a scientific agenda to begin with the whole IMoSEB [...] This came more as a consequence of *DIVERSITAS* trying to reinvent itself as an organization that wasn't very influential. At that time they got the backing of France and UNESCO had some ambitions to boost something like that. So it was more driven by a scientific community agenda than it was by a political process.
>
> (Global Biodiversity Information Facility [GBIF], Secretary, Interview, June 2010, in Busan/Rep. of Korea at the third IPBES meeting)

18 For a report of the meeting see (IMoSEB Report, 2008: 59).

19 In particular, France and Germany are named.

20 www.publications.parliament.uk/pa/cm200607/cmselect/cmenvaud/77/77we05.htm

21

> These are things which could be enhanced within the CBD. SBSTTA could be more effective. We could remove much of the policy issues currently being held by SBSSTA and either create a subsidiary body on implementation (SBI) under the CBD. Brazil has proposed that since COP4 to establish a SBI under the CBD. But some countries argue that that would mean that more money will be spent on the CBD budget. So, some countries blocked the discussion.
>
> (Member of the Brazilian Delegation to CBD COP, Interview, 20 May 2009 in Bonn/Germany at COP 9 of the CBD)

22 "[t]akes note of the outcomes of the consultative process towards an international mechanism of scientific expertise on biodiversity (IMoSEB) (UNEP/CBD/COP/9/INF/34)" and

> *welcomes* the agreement of the Executive Director of the United Nations Environment Programme to convene an *ad hoc* open-ended intergovernmental multi-stakeholder meeting to consider establishing an efficient international science-policy interface on biodiversity, ecosystem services and human well-being, [...] and *invites* Parties to ensure that appropriate science and policy experts are made available to attend, and also encourages the participation of experts from various regions and disciplines.
>
> (COP Decision IX/15)

23 www.cbd.int/doc/meetings/cop-bureau/cop-pres-msg-08-11-en.pdf
24 For a detailed list of attending countries, UN organizations, conventions and agencies, NGOs and other organizations, see (UNEP, 2008: 3ff).
25

> So we had some people coming to that meeting, who didn't know about the MA follow-up process, didn't know about IMoSEB, and yet the meeting was framed up as a decision meeting to say let's start and launch a new body. You had at that meeting people, who had never heard of anything right to the other hand of the spectrum, which means people that had been involved already for three years – working very hard. So it was a frustrating meeting, because it was the first time where everyone got together in the same room and there was just a very different familiarity of what the problem was, the potential issues were.
>
> (Member of the Canadian Delegation to IPBES, Interview, June 2010 in Busan/Rep. of Korea at the third IPBES meeting)

26

> Half way through the first meeting, I knew that it was going to be difficult. We understood the concept generically, in general. But not well enough to support it. So that's actually it. At that first meeting in Putrajaya, I called for the gap analysis to make sure that all the countries would be at the same page and have a common understanding of what we would be trying to accomplish. At that point I knew it was going to be critical and that it was going to take more than one and maybe more than two meetings.
>
> (Member of the Canadian Delegation to IPBES, Interview, June 2010 in Busan/Rep. of Korea at the third IPBES meeting)

27 Robert Watson – who was at that time Chief Scientific Advisor for the UK Department of Environment, Food and Rural Affairs (DEFRA) and Chair of Environmental Sciences at the University of East Anglia – did not only play an important role in bringing together the very diverging views of the meeting participants, but also in providing information on, for instance, the IPCC procedures and the development of global assessments. Compared to the Chair and other Vice-Chairs, he was much more dominant and influential, providing the plenary with summaries and propositions throughout the meeting. Robert Watson was also IPCC Chair from 1997 to 2002 and Board Co-Chair for the Millennium Ecosystem Assessment from 2000 to 2005. He received several honors and prizes, *inter alia* he received the Champions of the Earth Award from UNEP in 2014.
28 The analysis was prepared by UNEP with the assistance of the UNEP World Conservation Monitoring Centre (UNEP-WCMC) and input from a wide range of governments, IGOs, NGOs and individuals (UNEP, 2009: 2). It consists of five

chapters and was *inter alia* based on an online peer review of the preliminary gap analysis from March to June 2009 to which 788 comments from 54 respondents (including 21 from governments) were received, which was fewer than expected "despite direct requests to governments and other stakeholders" (UNEP, 2009: 8).

29 An interview with a scientist and advisor to CBD COPs reveals a rather sceptical view:

> If everyone would have a logical fashion they would not insist to this, because it doesn't make any sense. But because different actors all see benefits for them to have an IPCC-like structure there is still this push happening even if it is the worst thing that could happen. And I don't think it will happen.

(UK Joint Nature Conservation Committee, Head of Unit, Interview, September 2009 in Visby/Sweden at the meeting of the European Platform for Biodiversity Research Strategy)

30 For detailed information see (UNEP/IPBES, 2010b: 2).

31 "As the meeting was convened in accordance with decision SS.XI/4 of the Governing Council, the rules of procedure of the Governing Council applied, *mutatis mutandis*, to its proceedings" (UNEP/IPBES, 2010b: 2).

32 From the perspective of one delegate from South Africa, determination of the function of IPBES and its institutional arrangements should at this stage of the negotiations be an intergovernmental process and the role of the scientific communities diminished:

> The third meeting now is about if we should establish it or not. Whether science is saying yes or saying no, is not important, it is more about governments. The nature of this meeting is different from the nature of the first two meetings. The first and second meetings were trying to clarify a lot of issues. Now it is more governance and the functions and the form, which may not be too much a science discussion *per se*, but more an administrative, governance, and legal discussion.
> (Member of the South Africain Delegation, Interview, June 2010 in Busan/Rep. of Korea at the third IPBES meeting)

33 Some delegates criticized the Chair for moving the agenda forward even though important issues remained unresolved.

34 1. Collaborate with existing initiatives on biodiversity and ecosystem services, including multilateral environmental agreements, United Nations bodies and networks of scientists and knowledge holders, to fill gaps and build upon their work, while avoiding duplication;
   2. Be scientifically independent and ensure credibility, relevance and legitimacy through the peer review of its work and transparency in its decision-making processes;
   3. Use clear, transparent and scientifically credible processes for the exchange, sharing and use of data, information and technologies from all relevant sources, including non-peer-reviewed literature, as appropriate;
   4. Recognize and respect the contribution of indigenous and local knowledge to the conservation and sustainable use of biodiversity and ecosystems;
   5. Provide policy-relevant information, but not policy-prescriptive advice, mindful of the respective mandates of the multilateral environmental agreements;
   6. Integrate capacity-building into all relevant aspects of its work according to priorities decided by the plenary;
   7. Recognize the unique biodiversity and scientific knowledge thereof within and among regions, and also recognize the need for the full and effective

participation of developing countries and for balanced regional representation and participation in its structure and work;

8. Take an interdisciplinary and multidisciplinary approach that incorporates all relevant disciplines, including social and natural sciences;
9. Recognize the need for gender equity in all relevant aspects of its work;
10. Address terrestrial, marine and inland water biodiversity and ecosystem services and their interactions;
11. Ensure the full use of national, subregional and regional assessments and knowledge, as appropriate.

35 Decision X/11 adopted by the Conference of the Parties to the Convention on Biological Diversity at its 10ᵗʰ meeting (October 2010).
36 Decision 185 EX/43 adopted by the United Nations Educational, Scientific and Cultural Organization Executive Board at its 185ᵗʰ Session (October 2010).
37 Resolution 65/162, adopted by the United Nations General Assembly at its 65ᵗʰ session (December 2010).
38 Resolution 14/2011. Intergovernmental Science-Policy Platform on Biodiversity and Ecosystem Services, adopted by the Food and Agriculture Organization of the United Nations at its 37ᵗʰ session (July 2011).
39 Conclusions of the 25ᵗʰ Session of the Animals Committee of the Convention on International Trade in Endangered Species of Wild Fauna and Flora (July 2011).
40 The reason for the lack of certainty relates to the language used in the decision of UNGA regarding IPBES (UNGA, 2011). UNGA "took note" of the decisions made within the framework of the meetings and conferences outlined above and

> requests the United Nations Environment Programme, without prejudice to the final institutional arrangements for the intergovernmental science-policy platform on biodiversity and ecosystem services and in consultation with all relevant organizations and bodies, in order to fully operationalize the platform, to convene a plenary meeting providing for the full and effective participation of all Member States, in particular representatives from developing countries, to determine modalities and institutional arrangements for the platform at the earliest opportunity.
>
> (UNGA, 2011: 5)

41 The reasons provided for this judgment are that the language used in the Busan Outcome ("recommends the establishment") and the resolution of the UNGA ("takes note of"); neither say that the platform was established nor do they imply that the platform should be established in order to be operationalized (UNEP/IPBES, 2011: 6).
42 Ibid.
43 Notes by A. Vadrot.
44 Initially the proposed title for this subsidiary body was "multidisciplinary scientific expert panel". It was very much opposed by Bolivia, China and Argentina. They argued that this would restrict the scope of the knowledge relevant to IPBES to scientific knowledge only.
45 Bolivia argued that IPBES should *inter alia* mention "respect for human rights, including the rights of indigenous peoples, and equity in the development of approaches to non-commoditization of ecosystem services and functions" (UNEP/IPBES 2012a: 4). As the latter point was not sufficiently highlighted in the final document, the representatives of Bolivia, Egypt, and Venezuela indicated that they had no mandate to be listed as signatories and that hence, "they should not be listed among the governments consenting to the resolution" (UNEP/IPBES, 2012a: 4).

46 "Access to and transfer of technology", including biotechnology, is an important issue covered by CBD in Article 16 of the convention. Paragraph two of Article 16 says that

> Access to and transfer of technology [...] to developing countries shall be provided and/or facilitated under fair and most favourable terms, including on concessional and preferential terms where mutually agreed, and, where necessary, in accordance with the financial mechanism established by Articles 20 and 21. In the case of technology subject to patents and other intellectual property rights, such access and transfer shall be provided on terms which recognize and are consistent with the adequate and effective protection of intellectual property rights.
>
> (CBD Article 16)

Conflicts around "technology transfer" in the framework of CBD were present since the beginning, "With the introduction of industrial patents, access to improved breeding material may be hampered, as prices for seed increase. Some fear that patents will place constraints on technology transfers in general" (Rosendal, 2001) According to Görg and Brand (2000), the conflicts in place relate to the very specific interrelation between conservation and technology, i.e. technical use. "In principle, possession of technology and natural resources is in inverse proportion to the interest in both these spheres in both north and south" (Sánchez & Juma, 1994). It must be borne in mind, however, that the situation is considerably more complex than this simple model suggests. Neither north nor south is a homogeneous bloc or corresponds in any way to the geographical facts, but rather to socioeconomic ones, if at all. Regarding both agriculture and the issue of "wild" biodiversity, conflicts of interest permeate the individual states and state groups (Görg & Brand, 2000).

47 A detailed analysis of the narratives is provided in Vadrot, 2014: 215–279.

References

Agrawal, A. 2005. *Environmentality: Technologies of Government and the Making of Subjects.* Durham NC: Duke University Press.

Amano, T. & Sutherland, W. J. 2013. Four Barriers to the Global Understanding of Biodiversity Conservation: Wealth, Language, Geographical Location and Security. *Proceedings of the Royal Society B: Biological Sciences*, 280, 20122649.

Barbault, R. & Le Duc, J.-L. 2005. *Biodiversity, Science and Governance: Proceedings of the International Conference: Biodiversity, Science and Governance*, Muséum national d'Histoire naturelle, Bonchamps: Barneoud.

Bernstein, S. 2013. Rio+20: Sustainable Development in a Time of Multilateral Decline. *Global Environmental Politics*, 13, 12–21.

Biermann, F. 2002. Institutions for Scientific Advice: Global Environmental Assessments and Their Influence in Developing Countries. *Global Governance*, 8, 195–219.

Biermann, F. & Bauer, S. 2004. Assessing the Effectiveness of Intergovernmental Organisations in International Environmental Politics. *Global Environmental Change*, 14(2), 189–193.

Brand, U. 2010a. *Globale Umweltpolitik und Internationalisierung des Staates: Biodiversitätspolitik aus strategisch-relationaler Perspektive*. Münster: Westfälisches Dampfboot.

Brand, U. 2010b. Sustainable Development and Ecological Modernization – The Limits to a Hegemonic Policy Knowledge. *Innovation: The European Journal of Social Science Research*, 23, 135–152.

Brand, U. & Vadrot, A. B. M. 2013. Epistemic Selectivities and the Valorisation of Nature: The Cases of the Nagoya Protocol and the Intergovernmental Science-Policy Platform for Biodiversity and Ecosystem Services. *Law, Environment and Development Journal*, 9, 202–220.

Briggs, S. V. & Knight, A. T. 2011. Science-Policy Interface: Scientific Input Limited. *Science*, 333, 696–697.

Brooks, T. M., Lamoreux, J. F. & Soberón, J. 2014. IPBES ≠ IPCC. *Trends in Ecology and Evolution*, 29, 543–545.

Carpenter, S. R., Pereira, H. M., Perrings, C., Reid, W. V., Sarukhan, J., Scholes, R. J., Whyte, A., Clark, W. C., Mooney, H. A., Agard, J., Capistrano, D., Defries, R. S., Díaz, S., Dietz, T., Duraiappah, A. K. & Oteng-Yeboah, A. 2009. Science for Managing Ecosystem Services: Beyond the Millennium Ecosystem Assessment. *Proceedings of the National Academy of Sciences of the United States of America*, 106, 1305–1312.

Cash, D. W., Clark, W. C., Alcock, F., Dickson, N. M., Eckley, N., Guston, D. H., Jäger, J. & Mitchell, R. B. 2003. Knowledge Systems for Sustainable Development. *Proceedings of the National Academy of Sciences of the United States of America*, 100, 8086–8091.

Di Castri, F. & Younès, T. 1996. *Biodiversity, Science and Development: Towards a New Partnership*. Wallingford: CAB International in association with the International Union of Biological Sciences.

ENB 2008a. Summary of the Ninth Conference of the Parties to the Convention on Biological Diversity: 19–30 May 2008. *Earth Negotiations Bulletin*, 9(452). Available at: www.iisd.ca/download/pdf/enb09452e.pdf (accessed 16 December 2013).

ENB 2008b. A Report of the Ad Hoc Intergovernmental Multi-Stakeholder Meeting on an Intergovernmental Science-Policy Platform for Biodiversity and Ecosystem Services (IPBES): 10–12 November 2008. *IPBES Bulletin*, 158(1). Available at: www.iisd.ca/download/pdf/sd/ymbvol158num1e.pdf (accessed 16 December 2013).

ENB 2010a. IPBES III Highlights Wednesday, 9 June 2010. *IPBES Bulletin*, 3, 158(9). Available at: www.iisd.ca/download/pdf/sd/ymbvol158num9e.pdf (accessed 16 December 2013).

ENB 2010b. A Summary Report of the Third Ad Hoc Intergovernmental and Multi-Stakeholder Meeting on an Intergovernmental Science-Policy Platform on Biodiversity and Ecosystem Services (IPBES): 7–11 June 2010. *IPBES Bulletin*, Final Issue, 158(11). Available at: www.iisd.ca/download/pdf/sd/ymbvol158num11e.pdf (accessed 16 December 2013).

ENB 2010c. Summary of the Tenth Conference of the Parties to the Convention on Biological Diversity: 18–29 October 2010. *Earth Negotiations Bulletin*, 9(544). Available at: www.iisd.ca/download/pdf/enb09544e.pdf (accessed 16 December 2013).

ENB 2011. First Session of the Plenary Meeting on the Intergovernmental Science-Policy Platform on Biodiversity and Ecosystem Services (IPBES): 3–7 October 2011. *Earth Negotiations Bulletin*, 16(90–94). Available at: www.iisd.ca/ipbes/sop1/compilatione.pdf (accessed 16 December 2013).

ENB 2012a. IPBES-2 Highlights, Wednesday, 18 April 2012. *Earth Negotiations Bulletin*, 16(101). Available at: www.iisd.ca/download/pdf/enb16101e.pdf (accessed 16 December 2013).

ENB 2012b. IPBES-2 Highlights, Friday, 20 April 2012. *Earth Negotiations Bulletin*, 16 (103). Available at: www.iisd.ca/download/pdf/enb16103e.pdf (accessed 16 December 2013).

Esguerra, A. 2015. Toward Two Narratives of Knowledge. *Innovation: The European Journal of Social Science Research*, 28, 3–10.

Görg, C. & Brand, U. 2000. Global Environmental Politics and Competition between Nation-States: On the Regulation of Biological Diversity. *Review of International Political Economy*, 7, 371–398.

Granjou, C., Mauz, I., Louvel, S. & Tournay, V. 2013. Assessing Nature? The Genesis of the Intergovernmental Platform on Biodiversity and Ecosystem Services (IPBES). *Science, Technology and Society*, 18, 9–27.

Guay, L. 2002. The Science and Policy of Global Biodiversity Protection. *In*: Le Prestre, P. (ed.) *Governing Global Biodiversity. The Evolution and Implementation of the Convention on Biological Diversity*. Aldershot, UK: Ashgate.

Guruswamy, L. D. 1998. The Convention on Biological Diversity: A Polemic. *In*: Guruswamy, L. D. & McNeely, J. A. (eds.) *Protection of Global Biodiversity: Converging Strategies*. Durham, NC, London: Duke University Press.

Harper, J. L. & Hawksworth, D. L. 1994. Biodiversity: Measurement and Estimation. Preface. *In*: Hawksworth, D. L. & Royal, S. (eds.) *Biodiversity: Measurement and Estimation*. London: Chapman & Hall.

Heywood, V. H. 1995. Global Biodiversity Assessment. *Biodiversity Letters*, 1, 193–193.

Hulme, M., Wittmer, H., Böschen, S., Bridgewater, P., Diaw, M. C., Fabre, P., Figueroa, A., Heong, K. L., Korn, H., Leemans, R., Lövbrand, E., Mahony, M., Hamid, M. N., Monfreda, C., Pielke, R., Jr, Settele, J., Winter, M., Vadrot, A. B. M., van den Hove, S., van der Sluijs, J. P., Beck, S., Görg, C., Hansjürgens, B., Hauck, J., Nesshöver, C., Paulsch, A. & Vandewalle, M. 2011. Science-policy Interface: Beyond Assessments. *Science*, 333, 697–698.

IMoSEBReport 2008. *Strengthening the Science-policy Interface on Biodiversity. Results of the Consultative Process towards an IMoSEB*. Authors: Babin, D., Thibon, M., Larigauderie, A., Guinard, S., Monfreda, C. & Brels, S. Consultative process towards an IMoSEB, Executive Secretariat. April 2008. Montpellier.

Jessop, B. O. B. 2004. Critical Semiotic Analysis and Cultural Political Economy. *Critical Discourse Studies*, 1, 159–174.

Jutro, P. R. 1993. Human Influence on Ecosystems: Dealing with Biodiversity. *In*: McDonnel, M. J. & Pickett, S.T.A. (eds.) *Humans as Components of Ecosystems*. Berlin: Springer-Verlag.

Kanie, N., Betsill, M. M., Zondervan, R., Biermann, F. & Young, O. R. 2012. A Charter Moment: Restructuring Governance for Sustainability. *Public Administration & Development*, 32, 292–304.

Koetz, T., Bridgewater, P., van den Hove, S. & Siebenhüner, B. 2008. The Role of the Subsidiary Body on Scientific, Technical and Technological Advice to the Convention on Biological Diversity as Science–policy Interface. *Environmental Science & Policy*, 11, 505–516.

Koetz, T., Farrell, K. N. & Bridgewater, P. 2012. Building Better Science-policy Interfaces for International Environmental Governance: Assessing Potential within the Intergovernmental Platform for Biodiversity and Ecosystem Services. *International Environmental Agreements: Politics, Law and Economics*, 12, 1–21.

Larigauderie, A. & Mooney, H. A. 2010. The Intergovernmental science-policy Platform on Biodiversity and Ecosystem Services: Moving a Step Closer to an

IPCC-like Mechanism for Biodiversity. *Current Opinion in Environmental Sustainability*, 2, 9–14.

Le Prestre, P. G. 2002. Governing Global Biodiversity: The Evolution and Implementation of the Convention on Biological Diversity. *Global Environmental Governance Series*. Aldershot: Ashgate.

Loreau, M., Oteng-Yeboah, A., Arroyo, M. T. K.Babin, D., Barbault, R., Donoghue, M., Gadgil, M., Häuser, C., Heip, C., Larigauderie, A., Ma, K., Mace, G., Mooney, H. A., Perrings, C., Raven, P., Sarukhan, J., Schrei, P., Scholes, R. J., Watson, R. T. 2006. Diversity without Representation. *Nature*, 442, 245–246.

Lovejoy, T. E. 1997. Biodiversity: What is it? *In*: Reaka-Kudla, M. L., Wilson, D. E. & Wilson, A. E. O. (eds.) *Biodiversity II. Understanding and Protecting Our Biological Resources*. Washington DC: Joseph Henry Press.

MA 2005a.*Ecosystems and Human Well-Being*. General Synthesis. Millennium Ecosystem Assessment.Washington, DC: Island Press.

Mooney, H. & Mace, G. 2009. Biodiversity Policy Challenges. *Science*, 325(5947): 1474.

McConnell, F. 1996. *The Biodiversity Convention: A Negotiating History: A Personal Account of Negotiating the United Nations Convention on Biological Diversity, and After*. London, Boston: Kluwer Law International.

Mcgraw, D. M. 2002. The Story of the Biodiversity Convention: From Negotiation to Implementation. *In*: Le Prestre, P. G. (ed.) *Governing Global Biodiversity: The Evolution and Implementation of the Convention on Biological Diversity*. Aldershot: Ashgate.

Nature. 2010. Wanted: An IPCC for Biodiversity. *Nature*, 465(7298): 525–526. Available at: www.nature.com/nature/journal/v465/n7298/full/465525a.html (accessed 12 February 2016).

Neves, K., Turnhout, E. & Lijster, D. E. B. 2014. "Measurementality" in Biodiversity Governance: Knowledge, Transparency, and the Intergovernmental Science–Policy Platform on Biodiversity and Ecosystem Services (IPBES). *Environment and Planning A*, 46, 581–597.

Norgaard, R. B. 2008. Finding Hope in the Millennium Ecosystem Assessment. *Conservation Biology*, 22, 862–869.

Norton, B. G. 2001. Conservation Biology and Environmental Values: Can there be a Universal Earth Ethic? In: C. Potvin, et al. (eds.), *Protecting Biological Diversity: Roles and Responsibilities*. Montreal: McGill-Queen's University Press.

Norton, B. G. & Hannon, B. 1997. Environmental Values: A Place-based Theory. *Environmental Ethics*, 19, 227–244.

Olembo, R. J. 1994Bridge-Building for Biodiversity. *In*: Di Castri, F. & T. Younès (eds.), *Biodiversity, Science and Development: Towards a New Partnership*. Paris: IUBS, 30–24.

Perrings, C. 1995. *Biodiversity Loss: Economic and Ecological Issues*. Cambridge: Cambridge University Press.

Perrings, C. 2010. The Economics of Biodiversity: The Evolving Agenda. *Environment and Development Economics*, 15, 721–746.

Perrings, C., Duraiappah, A., Larigauderie, A. & Mooney, H. 2011. The Biodiversity and Ecosystem Services Science-policy Interface. *Science*, 331, 1139–1140.

Reaka-Kudla, M. L., Wilson, D. E., Wilson, O.E. 1997. *Biodiversity II: Understanding and Protecting our Biological Resources*. Washington, DC: Joseph Henry Press.

Reid, W. V. 2006. *Millennium Ecosystem Assessment (MA): Survey of Initial Impacts*. Washington, DC: MA Technical Assessment Volumes.

Rosendal, G. K. 2001. Impacts of Overlapping International Regimes: The Case of Biodiversity. *Global Governance*, 7, 95–117.

Sánchez, V. & Juma, C. 1994. Biodiplomacy: Genetic Resources and International Relations. *ACTS Environmental Policy Series*, no. 4. Nairobi, Kenya: ACTS Press, African Centre for Technology Studies.

Solbrig, O. T. 1991. *From Genes to Ecosystems: A Research Agenda for Biodiversity*. Paris: IUBS.

Swanson, T. 1999. Why Is There a Biodiversity Convention? The International Interest in Centralized Development Planning. *International Affairs (Royal Institute of International Affairs 1944)*, 75, 307–331.

Takacs, D. 1996. *The Idea of Biodiversity: Philosophies of Paradise*. Baltimore, MD: Johns Hopkins University Press.

UNEP 2009. Preliminary Gap Analysis for the Purpose of Facilitating the Discussions on How to Strengthen the Science-policy Interface. Governing Council of the UNEP, UNEP/GC.25/INF/30.

UNEP/GCSS 2009. Intergovernmental Science-policy Platform on Biodiversity and Ecosystem Services, Report of the Executive Director. Eleventh special session of the Governing Council/Global Ministerial Environment Forum, Bali, Indonesia, 24–26 February 2010. UNEP/GCSS.XI/7. Available at: http://www.unep.org/gc/gcss-x/download.asp?ID=1206(accessed 20 January 2014).

UNEP/IPBES 2008. Report of the First ad hoc Intergovernmental and Multi-Stakeholder Meeting on an Intergovernmental Science- Policy Platform for Biodiversity and Ecosystem Services. UNEP/IPBES/1/6.

UNEP/IPBES 2009. Executive Summary: Science-policy Interface on Biodiversity and Ecosystem Services – Gap Analysis. Second ad hoc Intergovernmental and Multi-stakeholder Meeting on an Intergovernmental Science-policy Platform on Biodiversity and Ecosystem Services. UNEP/IPBES/2/2, 3 August 2009. Available at: www.ipbes.net/ meetings/Documents/UNEP_IPBES_2_2_Executive_Summary_rev_clean_en.pdf (accessed 20 January 2014).

UNEP/IPBES 2010a. Millennium Ecosystem Assessment: Experiences and Impacts. Third ad hoc Intergovernmental and Multi-stakeholder Meeting on an Intergovernmental Science-policy Platform on Biodiversity and Ecosystem Services. UNEP/IPBES/3/INF/12, 7 June 2010. Available at: http://ipbes.net/images/stories/documents/Appeal_IPBES3.pdf (accessed 20 January 2014)

UNEP/IPBES 2010b. Report of the Third ad hoc Intergovernmental and Multi-stakeholder Meeting on an Intergovernmental Science-policy Platform on Biodiversity and Ecosystem Services. UNEP/IPBES/3/3, 11 June 2010. Available at: www.ipbes.net/previousipbes-meetings/3rd-meeting-on-ipbes.html" (accessed 20 January 2014).

UNEP/IPBES 2011. Legal Advice of the Office of Legal Affairs of the United Nations Concerning Certain Legal Issues Pertaining to an Intergovernmental Science Policy Platform on Biodiversity and Ecosystem Services: Note by the Assistant Secretary-General for Legal Affairs to the Chair of the Plenary Meeting. UNEP/IPBES.MI/1/INF/14.

UNEP/IPBES 2012a. Executive Summaries of the Offers Submitted by the Governments of France, Germany, India, Kenya and the Republic of Korea to Provide the Physical Location of the Secretariat of an IPBES. UNEP/IPBES.MI/2/5. Available at: www.ipbes.net/ component/docman/doc_download/694-ipbes-mi-2–5-englishfin.html?Itemid=58 (accessed 20 January 2014).

UNEP/IPBES 2012b. Report of the Second Session of the Plenary Meeting to Determine Modalities and Institutional Arrangements for an Intergovernmental Science-policy Platform on Biodiversity and Ecosystem Services. UNEP/IPBES.MI/2/9, 19 May 2012. Available from: www.ipbes.net/component/docman/doc_download/987-fina lreport-panamaenglish.html?Itemid=159 (accessed 20 January 2014).

UNGA 2005. World Summit Outcome Resolution. A/RES/60/1.

UNGA 2011. Resolutions, 64th Session: General Assembly of the United Nations. Available from: http://www.un.org/en/ga/64/resolutions.shtml (accessed 6 June 2016).

Vadrot, A.B.M. 2011. Biodiversity and Society. Why Should Social Sciences Have a Say? – An Editorial. *Innovation: The European Journal for Social Science Research*, 24(3): 211–216.

Vadrot, A. B. M. 2014a. The Epistemic and Strategic Dimension of the Establishment of the IPBES: "Epistemic Selectivities" at Work. *Innovation: The European Journal of Social Science Research*, 27(4), 361–378.

Vadrot, A. B. M. 2014b. *The Politics of Knowledge and Global Biodiversity*. London and New York: Routledge.

Vadrot, A. B. M., Heumesser, C. & Ritzberger, M. 2010. Wissenschaft als Instrument und Akteur. Die Diskussion um ein Science-Policy Interface. *In*: Brand, U. (ed.) *Globale Umweltpolitik und Internationalisierung des Staates. Biodiversität aus strategisch-relationaler Perspektive*. Münster: Westfälisches Dampfboot.

Vohland, K. & Nadim, T. 2015. Ensuring the Success of IPBES: Between Interface, Market Place and Parliament. *Philosophical Transactions of the Royal Society B: Biological Sciences*, 370, 20140012–20140016.

Wells, M. P., Grossman, D. & Navajas, H. 2006. *Terminal Evaluation of the UNEP/ GEF Project "Millennium Ecosystem Assessment"*. United Nation Environment Programme, Evaluation and Oversight Unit.

Wilson, E. O. 1988. *BioDiversity*. National Academy of Sciences, Washington DC: National Academy Press.

4 IPBES mandate and governance

Denis Pesche, Guillaume Futhazar and Sandrine Maljean-Dubois

The Intergovernmental Science-Policy Platform on Biodiversity and Ecosystem Services (IPBES) faces various challenges, as highlighted in chapters 2 and 3. How can biodiversity regime complex initiatives be coordinated and integrated to improve the science-policy interface? How can an intersubjective understanding of the contents of biodiversity science be achieved in a setting of different knowledge systems, competing policy priorities, and rival agendas? How can intergovernmental dynamics be combined at different scales in selecting and assembling knowledge?

The Platform design is the result of a relatively long and laborious process of international negotiations (2008 to 2012) in the context of a biodiversity regime complex characterized by a large number of international conventions of various scopes (both general and thematic) and at all scales (global, regional and even bilateral) (see chapter 2). IPBES is governed by an array of rules and procedures which have been developed throughout this slow genesis (2008 to 2012) and regularly enhanced in every plenary session. These rules are related to the decision-making mechanism but also to the working procedure to provide a framework for and implement scientific assessments and other activities linked to the IPBES work program.

The aim of this chapter is to provide a broad overview of the dynamics of governing IPBES on two aspects, i.e. institutional functioning and its core business, namely the production of assessments on biodiversity issues. When striving to understand the dynamics of IPBES, we focus on some aspects that have been controversial among the participants, including the question of the legal status of IPBES.

The plenary sessions highlight the IPBES functioning process.[1] Generally, an IPBES Plenary agenda is structured as follows. Firstly, the organizational matters aim to finalize the agenda, to specify the membership of the Platform and validate the presence of observers. The representatives' credentials enable an evaluation of the exact number of members represented at the Plenary. Secondly, the work program and then the financial and budgetary arrangements for the Platform are discussed in the Plenary sessions. Further, the Platform operating rules and procedures are discussed, followed by debates on communications and stakeholder engagement strategies. Then the members discuss the institutional arrangements, including United Nations collaborative partnership arrangements for the work of the Platform and its Secretariat. Finally, the provisional agenda, date and venue of future plenaries are tabled, followed by the adoption of decisions and reports of the session.

Throughout this chapter, we strive to grasp the main elements of this conventional IPBES Plenary structure. This chapter will be a useful guide for readers who have not participated in an IPBES Plenary, giving them some insight into the Plenary process.

We have grouped these elements in two main parts:

1   What is the legal status of IPBES and how is it governed? This part focuses mainly on the decision-making process, including the organizational matters, rules of procedure of the Plenary, the Bureau and the multidisciplinary expert panel (MEP). We also include the institutional arrangements and financial and budgetary arrangements that shed light on the IPBES governing dynamics.
2   What is IPBES doing? This part focuses on IPBES activities, including the IPBES work program and some elements of its budgetary dimension that highlight the priorities of the Platform.

Through flashbacks, we reconsider some elements of debates which have taken place in IPBES preparatory meetings since 2008 in order to underline the possible diversity of options, which were then discussed, finally leading to an agreement. We also try to underline the challenges and potential controversies linked to some rules and procedures. Most of the rules were defined at the Panama meeting (April 2012), where 94 consenting governments formally established "an independent intergovernmental body to be known as the Intergovernmental Science-Policy Platform on Biodiversity and Ecosystem Services" (UNEP, 2012c).[2] Since 2012, the members have been meeting almost on a yearly basis and the Secretariat is based in Bonn (Germany).

IPBES status and governance

IPBES founding: a first that is not "the" first

As shown in the chapter on the IPBES history, the Busan meeting, held in 2010, ended with the agreement that IPBES "should be established". It would now be interesting to look more closely into the legal aspect of this emergence process. The meeting detailed the exact mission of the Platform and agreed on a number of principles that would constitute the foundation of its future structure (UNEP, 2010). The final meeting document was submitted at the 65th session of the United Nations General Assembly (UNGA). UNGA adopted a resolution calling for UNEP to organize a Plenary meeting in order to operationalize the Platform as soon as possible (UNEP, 2011a).

This resolution was taken into account by the UNEP governing council during its 26th session in February 2011, which:

> Decides, based on the request by the General Assembly in its resolution 65/162 of 20 December 2010, without prejudice to the final institutional arrangements for the intergovernmental science-policy platform on biodiversity and ecosystem services and in consultation with all relevant organizations and bodies, in order fully to operationalize the platform, to convene a Plenary meeting providing for the full and effective participation of all member states, in particular representatives from developing countries, to determine modalities and institutional arrangements for the platform at the earliest opportunity.
>
> (UNEP, 2011b)

But did UNGA really establish IPBES? Considering the fact that the Platform Plenary was scheduled to take place in October 2011, one might rightfully think that it did. However, in this existential limbo, negotiators felt it was necessary to ask the UN Office of Legal Affairs (OLA) this crucial question on this matter, but the office came to a different conclusion: "the General Assembly, by merely taking note of the relevant decisions in paragraph 17 of resolution 65/162 of 20 December 2010, did not express approval or disapproval of the arrangement outlined therein, and accordingly did not establish the platform as a United Nations body" (UNEP, 2011c). It also added that the Busan Outcome provided declarations on the Platform that were recommendations but did not make the decision to establish it.

Thus, the mandate of the October 2011 meeting, which was referred to as a "plenary", was to determine the modalities and institutional arrangements of the Platform. OLA also underlined that it did not have the mandate to operationalize the Platform, as it was not truly the first meeting of IPBES. If so, then what was the exact status of the ongoing Plenary? Was it part of an official intergovernmental process? According to OLA it was actually an intergovernmental process convened following a request by UNGA and in conformity with a decision from the UNEP governing council that had defined its relevant mandate. But this situation raises numerous questions. If the Platform was not yet established, who had the responsibility for establishing it? And under what conditions? Would the IPBES be within or outside of the UN system? Would it have a legal status?

At the end of this 2011 "pseudo plenary", several options were available for the establishment of the Platform. It could be established by the ongoing meeting that could then turn into a Plenary. According to the opinion of OLA, it could also be established by the relevant organizations through a decision of their secretariat heads. In this situation, the Platform could work in an autonomous fashion and be independent of the decision-making process of these organizations, while still remaining in their institutional framework as an intergovernmental body. Other options were also considered, such as the creation by UNGA or the UNEP governing body or even by converging decisions by the executive heads of UNEP, UNDP, FAO and UNESCO.[3] According to the latter hypothesis, IPBES would then be similar to the IPCC in its creation. These different options are not negligible as each of them has repercussions on the independence of the Platform to be established.

Finally, a year later, on April 21, 2012, IPBES was formally established by a resolution during the second Plenary meeting in Panama. At the time of the resolution's adoption, representatives from Bolivia, Egypt and Venezuela indicated that their governments were not to be listed as supporting this resolution. Most notably, they expressed dissatisfaction on the fact that a UN status had not been attributed to the newly established Platform from the outset.

As though to stress their intention to maintain as much control as possible over the Platform, the 2012 resolution specifically states that, "any future decisions of the Platform have a legally non-binding nature". Also, some of its members are prompted to recall that the IPBES has no international legal capacity thus limiting its autonomy to act outside of what its members have formally agreed to (ENB, 2015).

However, this tendency to promote state sovereignty at all costs has some limits – the resolution and the future decisions will inevitably have a "normative strength" (Thibierge, 2009) and measurable practical consequences, e.g. decisions on the geographical location of the Platform or the way it operates (Brunnee, 2002). Of course, States will be able to leave the Platform as freely as they have joined it, but the distinction between binding and non-binding, hard and soft, is not a crucial concern when considering the repercussions of the different instruments used. Most of them, despite having an uncertain normativity, are nevertheless being applied on a daily basis without the issue of their legal nature ever being raised. On the other hand, many conventional or customary obligations are simply not enforced. Ultimately, "so long as the stage of mutual interest continues relatively undisturbed, the legal aspects of the relationship may seem of minor importance" (Lachs, 1972). Could legitimacy be the most important factor in the "compliance pull" (Boyle and Chinkin, 2007) of many customary and conventional obligations? This statement does not imply that the procedures and processes of normative creation are irrelevant. On the contrary, the more open, transparent and inclusive they are, the more these norms will satisfy legitimacy criteria (Brunnee, 2002). This is a crucial stake for the Platform.

Plenary sessions – the main power arena for members

Since its official launch during the Panama meeting (2012), the Plenary has become the Platform's decision-making body, consisting of all Platform members. The latter are UN state members who have expressed their intent to be Platform members.

The main IPBES rules of procedure were agreed at this 2012 meeting and amended at the first IPBES Plenary in Bonn (IPBES, 2013b). Those rules were based on the relevant rules of procedure of the UNEP Governing Council (UNEP, 1988). The Platform members agreed that the meeting would make its decisions by consensus or, if not possible, by a two-thirds vote of present and voting Platform members (rule 36).[4]

The issue of membership of regional economic integration organizations remains under discussion. In the Busan meeting (2010), it was agreed that the Platform would be open to regional economic integration organizations – a provision that concerns the European Union, which is part of numerous multilateral conventions, and international organizations (such as WTO) under this status. However, during the Panama meeting, the USA, followed by the African group and GRULAC[5] opposed the fact that EU could access membership. The voting rules and the possibility for the European Commission to have electoral functions within the different IPBES bodies through one of its representatives caused some internal tension between the Commission and the member states. In the end, the Panama Resolution stated that, "the Platform is open to state members of the United Nations, who may become members by expressing their intent to do so". For the time being, the European Union only has an observer status,[6] and it seems unlikely that this situation will change in the foreseeable future (see rule 5). The last Plenary, held in 2015 in Bonn, did not even address this issue, suggesting that there was a general understanding that an agreement was still impossible to reach at this point in the process (IPBES, 2015b).

Platform members are represented by a delegation consisting of a head of delegation, other accredited representatives, alternate representatives and advisers. The credentials of representatives of Platform members assigned by or on behalf of a head of state or a foreign affairs minister must be submitted to the Secretariat and approved by the Plenary (rule 11 to 14). The presence of a majority of Platform members participating in the session is required for any decision to be made (rule 35).

Opening the process – observers and/or stakeholders

After an early stage of the negotiation process generally open to all stakeholders, would IPBES end up having a usual intergovernmental composition, or would it be more "modern", participative, open to private actors such as NGOs, scientists and even businesses and industry?

Non-state actors have actually been integrated in the IPBES process since the outset. In the first IPBES preparatory meeting (Malaysia, 2008), a strict intergovernmental option was presented with a second option more open to other stakeholders:

> The Platform is open to all states that are members of the United Nations or specialized agencies, in addition to relevant organizations and stakeholders. Each government has one vote. The relevant organizations and stakeholders also have qualified votes, the total number of

which should not exceed the total number of votes of participating governments.

(UNEP/IPBES/1/4)

This meeting did not make any decisions on that aspect and most of the participants expressed their preference for an intergovernmental platform (UNEP/IPBES/1/6). This was the choice at the Busan meeting (2010), confirming a quite common approach, calling for the participation of UN states and regional economic integration organizations. The document states that "Intergovernmental organizations and other relevant stakeholders should participate in the plenary as observers" and would therefore have no role in the formal decision-making process (UNEP, 2012a).

IPBES communicated on this will to provide open and easy observer access to the Plenary, with particular reference to the inclusion of intergovernmental organizations, non-governmental organizations, indigenous peoples and local communities. Some specific rules were established to regulate this openness.

Firstly, IPBES defines who are potential observers: "observers" are "any state not a member of the Platform and any body, organization or agency, whether national or international, governmental, intergovernmental or nongovernmental, including organizations of and representatives of indigenous peoples and local communities, which is qualified in matters covered by the Platform" (UNEP, 2012b).

Secondly, IPBES defines the procedure for admitting an observer. The question reached a deadlock during the first Platform Plenary in Bonn on the question of consensus or vote for the admission of new observers (ENB, 2013a). An interim procedure for the admission of new observers has thus been applied since then (IPBES, 2013a). This interim procedure reflects the will of some Platform member states to have complete control over the process by regulating the participation of non-state actors as much as possible. Basically, the current procedure automatically grants a Plenary observer status to any institution that has attended the previous session. Any actor wishing to be an observer to an upcoming Plenary has to send a formal request to the Platform Secretariat. The request is then scrutinized by the Platform Bureau, which then puts forward its observations to the Plenary on the admission of the new observer. However, a member state can oppose the admission of an observer. If such an opposition were to happen, it could only be overruled by a two-thirds majority of present and voting Platform members. Considering the danger of such a situation in a process that strives to promote consensus,[7] it can be argued that this interim procedure could ultimately lead to a situation where one state could issue a discretionary opposition to the participation of an observer.

Currently, there are two opposing views at each Plenary on the observer issue. China requested the incorporation of a consensus requirement in the rule of procedure from the outset. The consensus requirement would then allow any state to regulate the admission of observers. EU states, on the other hand, have called for the establishment of a voting procedure where the admission of an observer could only be rejected by one third of the members (ENB, 2015). This would prevent any excessive exclusion of observers. However, these two options are radically opposed and it has been acknowledged within the Plenary that these "strong views" (Ibid.) have not changed throughout the negotiation process. Consequently, each Plenary has eluded the problem by inviting the next one to deal with the issue. Ultimately, it is very probable that the interim procedure will be perpetuated even for IPBES-5 (in 2017). Indeed, it seems unlikely that this sensitive issue will be addressed in a definitive way considering the heavy workload of the upcoming fourth Plenary where the first reports are due to be discussed.

Since 2013, a stakeholder engagement strategy has been drawn up, closely linked to the IPBES Communication and Outreach Strategy. In fact, the standard corporate communication of IPBES *per se* (dissemination of Platform products) is enhanced by the aim to mobilize a large array of stakeholders to support implementation of the IPBES work program. Stakeholders and observers claim to be highly involved in the IPBES process. They regularly highlight the lack of adequate funding to support stakeholder engagement and the stakeholder engagement strategy, and dependence on in-kind contributions to participate as stakeholders. The stakeholder group requested that IPBES estimate the in-kind contributions to the work plan including, for example, the cost of expert time in preparing assessments. However, stakeholders are not a group *per se* nor a community (see chapter on stakeholders). In the Plenary session, a stakeholder is not allowed to be accepted as an observer.

Bureau – a political balance

The question of subsidiary bodies was identified early in the 2008 meeting (UNEP, 2008) and strongly discussed during a further meeting without reaching a consensus (Nairobi, 2011). Two options were tabled – a first one with a single subsidiary body, including an expanded Bureau and a second option with a dual structure, an executive body plus a scientific advisory group. The first option was defended by the EU, Norway, Mexico, Turkey, *inter alia*, arguing for a more coherent and less cumbersome body.[8] The second option, advocated by Japan, Brazil, USA, China, Indonesia, *inter alia*, argued that "this would facilitate efficiency in administration and foster the independence of the scientific functions" (ENB, 2012). Finally, the Panama meeting decided on this dual option. The IPBES Plenary thus has two subsidiary bodies, i.e. a Bureau and a multidisciplinary expert panel (MEP). Much of the discussion at this meeting focused on MEP, about its strictly or not scientific character and the nomination mechanism, based on UN regions or more biogeographical criteria.[9] Finally, the first Plenary (Bonn, 2013) established the first Bureau and an interim MEP.

The Plenary Bureau includes the Chair, four Vice-Chairs, and five other officers. Each UN region is represented by two Bureau officers.[10] Candidates for the Bureau are proposed by governments for nomination by UN regions and elected by the Plenary for a three-year mandate. The Chair and four Vice-Chairs are selected with due consideration to scientific and technical expertise. The Chair "will be rotated among the five United Nations regions every three years without the possibility of re-election as Chair" (rule 15). "The Bureau meets regularly "to advise the Chair and the Secretariat on the conduct of business of the Plenary and its subsidiary bodies" (rule 16).

Some of those personalities have been quite involved in the process since the beginning, e.g. R.T. Watson was Chair (2009, 2011, 2012) and Vice-Chair (2008, 2010) in the previous IPBES meetings, while A. Oteng-Yeboah was Vice-Chair (2008, 2009, 2010), S. Barudanovic was Vice-Chair (2011, 2012), A.H. Zakri was Vice-Chair (2009) and S. Thomas Vice-Chair (2010). A few of them had jointly been involved in the Millennium Ecosystem Assessment (R.T. Watson, A. Oteng-Yeboah and A. H. Zakri). Only R.T Watson is not a biodiversity specialist, but rather a specialist in managing global environmental assessments.[11]

The current Chair, Abdul Hamid Zakri, was elected during IPBES-1 but an internal agreement specified that the Vice-Chair of the Western Europe and Other group would become the Bureau Chair for the next term, namely Robert Watson.

The Bureau Chair presides over Plenary meetings. In each Plenary, members can establish contact groups (CGs) to assist in the consensus forging process (CBD, 2010). In IPBES-2, three CGs were trained on the Platform work program, conceptual framework, budget and financial arrangements, and

Table 4.1 Chairs and Vice-Chairs elected for the first mandate (2013–2015)

		2013/2014	2014/2015
Africa	Mr. Alfred Apau Oteng-Yeboah (Ghana)	Vice-Chair	
Eastern Europe	Mr. Sergey Trepelkov (Russian Federation)	Vice-Chair	
	Ms. Senka Barudanović (Bosnia and Herzegovina)		Vice-Chair
Asia	Mr. Abdul Hamid Zakri (Malaysia)	Chair	
Latin America and Caribbean	Mr. Leonel Sierralta (Chile)		Vice-Chair
	Mr. Spencer Thomas (Grenada)	Vice-Chair	
Western Europe and Other	Mr. Robert T. Watson (United Kingdom)	Vice-Chair	

policies and procedures. CG meetings are organized so as to avoid overlap with Plenary meetings. There is usually a lot of discussion in those informal meetings. The groups are open to observers with some exceptions, as in IPBES-2 for the budget group: "some participants expressed the view that observers have a strong case to be involved in budget discussions, noting that many of them have provided in-kind contributions to the Platform and are key to its success" (ENB, 2013b). Other informal meetings are organized around the official plenary schedule, *inter alia*, regional meetings and stakeholders' meetings.

Multidisciplinary Expert Panel – in search of credibility

The second Plenary subsidiary body, the Multidisciplinary Expert Panel (MEP), involved 20 participants: five nominated by each of the five UN regions. This governance aspect was discussed in depth during the first Plenary in Bonn (2013). For many delegations, the independence of MEP and its focus on science issues was a key point for the credibility of IPBES. The MEP embodies the IPBES willingness for independence of the mobilized expertise. The MEP "will carry out the scientific and technical functions agreed by the Plenary" (rule 24). And candidates are elected by consensus in the Plenary for a three-year mandate with a possibility of re-election for one consecutive term. The MEP composition is also based on UN region equilibrium: "Candidates for the Panel are to be proposed by members [and observers] of the Platform for nomination by regions and election by the Plenary. (...) Taking into account disciplinary and gender balance, each region will nominate five candidates for membership to the Panel (rule 26)". This question of regional structuration of the MEP was a point of discussion and some countries like Bolivia, Mexico, Colombia, Brazil and Japan advocate a regional network structure for the MEP but, finally, the first Plenary in Bonn decided that "the members of the Panel are elected for their personal expertise and are not intended to represent any particular region (rule 25)". The notion of multidisciplinarity was defined in a *sui generis* manner:

> multidisciplinarity connotes an approach that crosses many disciplinary boundaries, knowledge systems and approaches to create a holistic approach, focusing on complex problems that require expertise across two or more disciplines. Multidisciplinarity arises when scientists (including natural and social scientists), policy and technical experts, natural resource managers, other relevant knowledge holders and users, interact in an open discussion and dialogue giving consideration to each perspective.[12]

The procedures adopted during the second Platform Plenary regarding the MEP promote regional, disciplinary and gender balance. Nevertheless, the nomination of MEP members remains the exclusive prerogative of states,[13] despite voices among stakeholders calling for a role in the nomination of

MEP members (ENB, 2013b).The question of criteria for MEP member selection was intensively discussed and the trade-off pooled a number of criteria: (a) scientific expertise in biodiversity and ecosystem services with regard to both natural and social sciences and traditional knowledge among Panel members; (b) scientific, technical or policy expertise and knowledge of the main elements of the Platform work program; (c) experience in communicating, promoting and incorporating science into policy development processes; (d) ability to work in international scientific and policy processes (rule 26). Linkage with other UN conventions is managed by cross participation: "The chairs of the scientific subsidiary bodies of the multilateral environmental agreements related to biodiversity and ecosystem services and the Intergovernmental Panel on Climate Change will be observers".

Links between the MEP and the Bureau were the focus of discussion with the aim of guaranteeing both the independence of expertise mobilization and coherence. Formally, the Panama Plenary decided that Bureau members would also be Panel observers during the interim period (2013–2014). The Bonn Plenary clarified the role of both subsidiary bodies: "The need for the Multidisciplinary Expert Panel to retain its independence (which was a crucial issue for the credibility of the Platform). Some of the delegates also highlighted the importance of the independence of the Platform. The point was also made that the Multidisciplinary Expert Panel must focus on scientific and technical issues, with political issues being addressed by the Bureau and Plenary. It was also noted that the MEP may wish to develop a code of practice for its members to ensure highest scientific integrity in its work" (UNEP, 2013). In fact, the Bureau and MEP try to meet at the same place and date, and Bureau members can attend MEP meetings as observers in order to improve the expected collaboration.

An interim period was decided to test the new IPBES organizational scheme. A mid-term "self-assessment" was conducted after three MEP and Bureau meetings, in June 2014. MEP and Bureau members highlighted the fact that they did not have sufficient social scientists and economists, nor marine specialists. They also stressed the difficulty of achieving a discipline and gender balance within UN regions (IPBES, 2015a). A social network study on MEP and Bureau members in 2014 showed that natural science (36/41) and male membership (32/41) dominated (Morin et al., 2015).

In 2015, the third Plenary led to the re-election of MEP members. Platform members tried to take the lessons learned from previous interim MEPs into account in order to constitute a more balanced Panel. Although some progress has been made in this respect, observers have underlined the fact that there is still room left for improvement (Montana and Borie, 2015).

Secretariat

During the inception phase (2008–2012), the Secretariat was supported by UNEP and based in Nairobi and then moved to Bonn (Germany) as decided at the Panama meeting (2012). The IPBES-1 Plenary (Bonn, 2013) "requests UNEP, UNESCO, FAO and UNDP to establish an institutional link with the Platform through a collaborative partnership arrangement for the work of IPBES and its Secretariat", and "UNEP to provide the Platform Secretariat, which will be solely accountable to the IPBES Plenary on policy and programmatic matters" (IPBES 1/EN Annex V). IPBES-2 (Antalya, 2013) established a collaborative partnership arrangement among the plenaries of IPBES and the United Nations Environment Programme (UNEP), the United Nations Educational, Scientific and Cultural Organization (UNESCO), the Food and Agriculture Organization of the United Nations (FAO) and the United Nations Development Programme (UNDP). This partnership aims to strengthen collaborations between them for implementation of the IPBES work program. The Secretariat staff is under the administrative responsibility of UNEP and other UN organizations provide staff support.[14]

The Secretariat has the following indicative administrative functions under the direction of the Plenary: organizing meetings, communication and outreach activities, support to budgetary functions and mobilization of financial resources, monitoring and evaluation of the Platform work (UNEP, 2012b). Recruitment for the Secretariat was carried out in 2014, with nine staff, and a further three more staff positions were requested in IPBES-3 (Bonn, 2015).

The IPBES-2 Plenary recognized that the needs for coordinating the work program deliverables would exceed the Secretariat's capacity and agreed that a cost effective way to provide the necessary additional technical support could be through different arrangements, such as technical support units (TSU), based on in-kind offers from governments and other stakeholders.

Since the IPBES-2 Plenary, three task forces have been created on: (1) capacity building (see chapter 9), (2) knowledge and data, and (3) indigenous and local knowledge (ILK) systems.[15] The capacity building and knowledge and data task forces are co-chaired by two Bureau members and include three MEP members, as well as nominated and selected experts. The ILK task force is co-chaired by two MEP members and includes two Bureau members, as well as nominated and selected experts. The Norwegian government accepted to host a technical unit for the capacity building group, while the South Korea government did the same for the knowledge and data task forces, and UNESCO for ILK. Those engagements and others (e.g. UNEP suggested hosting the scoping meeting for the regional assessments) illustrate the multistakeholder dynamics in the implementation of the work program. Then, "technical support units had been established for all three task forces and for the thematic assessment on scenario analysis and modelling, and other technical support arrangements had been made for the assessment on pollination and pollinators associated with food production and for supporting the delivery of regional and subregional assessments" (IPBES, 2015b).

This also led to a great challenge for the Secretariat, the Bureau and the MEP to coordinate such disseminated combinations of human and financial resources.

Platform financial and budgetary arrangements

Since its inception in 2012, "a core trust fund to be allocated by the Plenary" was to be established "to receive voluntary contributions from Governments, as well as from United Nations bodies, the Global Environment Facility, other intergovernmental organizations and other stakeholders such as the private sector and foundations, on the understanding that such funding will come without conditionalities, will not orient the work of the Platform and cannot be earmarked for specific activities" (IPBES Decision 2/7).

Three years later,[16] the third Plenary conducted a first assessment of the IPBES funding process. The Platform received cash and in-kind contributions. Cash contributions fueled the trust fund and represented US$2,236,000 in 2012, US$4,277,000 in 2013 and US$13,621,000 in 2014. The main contributing countries were Norway (41%), Germany (24%), United Kingdom (11%), USA (7%) and four other countries, i.e. Netherlands, Japan, France and Sweden (2–3%). In-kind contributions are mainly technical staff (71%) and meetings facilities (28%) for an estimated amount of US$2,470,000. The main in-kind contributors were Germany, Norway, China, Netherlands and Korea, but also UN organizations (UNEP, UNDP, UNESCO and FAO) and IUCN.[17]

The Platform expenses amounted to US$2,327,018 for the 2013 financial year and US$3,247,838 for 2014.[18] The 2013/2014 variation was mainly due to implementation of the work program (+US$1.6 M) and to staff recruitment within the Secretariat after the arrival of the Platform Executive Secretary in February 2014. An estimated revised budget for 2015 was established at US $9,526,779, including the work program contribution (US$5.5 M) and staff recruitment (US$1.6 M). Those budget increases reflect the sudden increase in IPBES activities after the first meetings that were mainly devoted to drawing up the rules and procedures.

The global estimated budget for the 2014–2018 work program is US$40.8 M. In late 2015, the Chair reminded members about the need to fuel the IPBES trust fund: US$13.6 M were still needed to reach the targeted budget.

Institutional arrangements

One of the main institutional questions at the outset of IPBES concerned links to establish with other United Nations bodies. On this issue, opinions were divided between a minimalistic link, where UNEP, UNESCO, FAO and UNDP would provide their support to the Platform, or a formal attachment to the UN family. This second option was favoured by African states who saw several safeguards in this proposition. Indeed, an UN body would be compelled to comply with UN principles and its objectives would have to be in line with those set within the UN Charter. The institutional framework of the different responsible bodies would have to be applied, most notably the relevant norms concerning the communication of data. The affiliated body could also use the institutional framework of the concerned organizations

(UNEP, 2011c). For an external body, these requirements would not apply. Moreover, the Platform – without institutional links with other UN bodies – would need additional funding for the supply of administrative services. Yet, the USA still preferred to apply specific rules of procedure to the Platform and was against such links. Even the establishment of the Platform in 2012 did not end this debate and it was left to the first Platform Plenary to decide on the relation between the Platform and the UN.

Discussions on this matter were heated during the first Plenary. Understandably so, as this issue raises the question of the role of other international institutions and UNEP, which is expected by the General Assembly to have a more important role in assisting in the decision-making process. In the end, it took two plenaries to draw up the procedural aspects of the relationship between the Platform and the other UN institutions. IPBES Decision 1/4 established the administrative and institutional arrangements of the Platform, where UNEP was requested to provide the Platform Secretariat (which is accountable to the Plenary) and UNDP, UNESCO and FAO were requested to establish institutional links, "through a collaborative partnership arrangement for the work of IPBES and its Secretariat".[19] IPBES Decision 2/8 further elaborated on this aspect and illustrated how the other institutions could assist the Platform in the implementation of its work program.[20] Finally, the framework of cooperation between IPBES and these other institutions is fairly straightforward – while UNEP is the only organization in charge of administering the Platform Secretariat, all organizations assist the Platform by participating in the implementation of its work program, sharing information, attending meetings or even providing staff.

This arrangement highlights the potential coordination role played by IPBES within the biodiversity complex regime (see chapter 2). One of the many questions regarding IPBES during its conceptualization concerned the exact potential relationship between the Convention on Biological Diversity (CBD) and the Platform. From the outset, the CBD Conference of Parties (COP) followed negotiations on IPBES very closely but remained more of a passive observer than an actor. Its very brief resolution X/11 simply called for UNGA to study the possibility of establishing an IPBES as soon as possible.[21] This decision highlighted the necessity for the Platform to address the requirements of CBD and strengthen SBSTTA rather than entering in an obvious competition with it. The decision also requested the CBD Secretariat, in collaboration with the Subsidiary Body on Scientific, Technical and Technological Advice (SBSTTA), to examine how CBD could make full and efficient use of IPBES by seeking complementarity and avoiding any duplication (UNEP, 2010).

This matter was and still is a focus of discussion within SBSTTA as the cooperation aspects between the Platform and the subsidiary body are yet to be clearly defined. For instance, the SBSTTA of the CBD issued a working document for its November 2015 meeting in which it discussed the different ways the Platform deliverables could fit into its own work program.[22] This document was produced following decision XII/25 of the CBD which specifically requested the SBSTTA to work on this issue.[23] The SBSTTA will be able to put forward suggestions to the Platform following the approval of those suggestions by the COP[24] (thus removing any possibility for the SBSTTA to directly access the Platform). The same rationale is being promoted among the other expert organs of the biodiversity regime complex. For example, the most recent meeting of the CITES Animal Committee (Tel Aviv, 2015) addressed the issue as to how the chairs of the different scientific bodies of the multilateral environmental agreements could be better coordinated in relation to MEP meetings.[25] These developments are more than welcome as the failure to develop partnerships with these bodies would hamper the Platform in efficiently performing some of its functions.

The whole process illustrates the complexity of global environment governance. This complexity and the necessity to constantly adapt to it is apparent in the reflexive design of the Platform, as illustrated in the Busan Outcome in which the members concluded that "the Platform's efficiency and effectiveness should be independently reviewed and evaluated on a periodic basis as decided by the Plenary, with adjustments to be made as necessary" (UNEP, 2012c). This demonstrates that the Platform is not conceptualized as being a rigid institution and has room for adaptation and evolution through iteration.

All of those aspects concerning members, the Bureau, the MEP, the budget and institutional arrangements constitute the institutional infrastructure of the Plenary and the intersessional activities.[26] Those elements of IPBES governance are oriented towards developing products and managing processes that are linked within the IPBES work program.

IPBES core activities – the work program

The Busan meeting (2010) clearly outlined the IPBES global mandate: "The Platform's objective is to strengthen the science-policy interface for biodiversity and ecosystem services for the conservation and sustainable use of biodiversity, long-term human wellbeing and sustainable development" (UNEP, 2012a) "Focusing on Government needs and based on priorities established by the Plenary, the Platform responds to requests from governments, including those conveyed to it by multilateral environmental agreements related to biodiversity and ecosystem services as determined by their respective governing bodies".

Four main functions were thus identified:

"The Platform identifies and prioritizes key scientific information needed for policy makers at appropriate scales and catalyses efforts to generate new knowledge by engaging in dialogue with key scientific organizations, policy makers and funding organizations, but should not directly undertake new research."

"The Platform performs regular and timely assessments of knowledge on biodiversity and ecosystem services and their interlinkages, which should include comprehensive global, regional and, as necessary, subregional assessments and thematic issues at appropriate scales and new topics identified by science and as decided upon by the Plenary."

"The Platform supports policy formulation and implementation by identifying policy relevant tools and methodologies (...) for decision makers."

"The Platform prioritizes key capacity-building needs to improve the science-policy interface at appropriate levels and then provides and calls for financial and other support for the highest-priority needs related directly to its activities."

(UNEP, 2012a)

Interestingly, the fact that this option includes four functions in the IPBES work program, rather than focusing solely on assessment as previous related initiatives have done, is presented as an innovative challenge in comparison to other science-policy interfaces. In particular, the Platform provides the necessary means for effective participation of developing countries by promoting capacity building. This matter was neglected by the IPCC when it was originally established. As some countries have more biodiversity than others, it is crucial to ensure that appropriate research can be conducted within their territories.

After the Panama meeting, the development of the initial IPBES work program was directly linked to those four functions.

The work program – four pillars

During the IPBES-1 meeting, concerning the preparation of the work program, the participants "identified a number of issues as being particularly important. These included: ensuring an appropriate balance across the four functions of the Platform; reducing potential bureaucracy in procedures; clarifying the respective roles of the Bureau and the Multidisciplinary Expert Panel in a number of activities; and the urgent need for further work on the conceptual framework" (IPBES-1, EN). The intersessional period between IPBES-1 and 2, was devoted to a specific activity around the conceptual framework (see chapter 7) which was adopted during the second Plenary in Antalya in December 2013 (IPBES Decision 2/4).

The work program is supposed to address requests by governments. How does IPBES manage the process for receiving and prioritizing requests put to the Platform? Those requests could be formulated by government members but also by United Nations bodies related to biodiversity and ecosystem services or other relevant stakeholders.[27] Requests, inputs and suggestions are to be presented to the Secretariat no later than six months prior. Then the Bureau and the MEP prioritize those proposals based on a previous evaluation conducted on the basis of various criteria, *inter alia*, the relevance, urgency, geographic scope and level of complexity, and prepare a report to be disseminated 12 weeks prior to the Plenary meeting for consideration and decision (IPBES Decision 1/3). As shown in chapter 2, governments remain clearly in the driver's seat in this process.

The IPBES work program adopted during the second Plenary session (IPBES Decision 2/5) is planned for five years (2014–2018) and structured in four objectives and 16 deliverables.

Objective 1 aims to, "Strengthen the capacity and knowledge foundations of the science-policy interface to implement key functions of the Platform" (IPBES, 2013a). For that, four deliverables were identified.[28] The first two are related to capacity-building needs with the aim of identifying capacity-building needs clearly linked to achieving the Platform work program (deliverable 1a) and to integrate the capacity-building needs into activities (1b). Those two deliverables will be produced with the support of a specific task force on capacity-building supported by UNDP (see chapter 9).

The third deliverable aims to produce procedures, approaches for participatory processes for working with indigenous and local knowledge systems developed (1c) with the creation of an *ad-hoc* task force hosted by UNESCO since July 2014 (see chapter 10). The fourth one aims to identify and prioritize knowledge and data needs for policy making (1d) and also led to the creation of an *ad-hoc* task force, which aims at, "helping to identify and prioritize key scientific information needed for policy makers at appropriate scales, and to catalyse efforts to generate new knowledge in dialogue with scientific organizations, policy makers and funding organizations".

The two next objectives are related to producing assessments on biodiversity and ecosystem services. *Objective 2* aims to, "Strengthen the science-policy interface on biodiversity and ecosystem services at and across subregional, regional and global levels". Three deliverables are identified. The first one is a guide on production and integration of assessments from and across all scales (2a) to address practical, procedural, conceptual and thematic aspects for undertaking assessments, taking different visions, approaches and knowledge systems into account. The second one is a set of regional and subregional assessments established through a regionally based scoping process (2b). The third one is a global assessment on biodiversity and ecosystem services (2c) to be prepared for the 2018 CBD COP.

Objective 3 aims to, "Strengthen the science-policy interface on biodiversity and ecosystem services with regard to thematic and methodological issues". Four deliverables are expected. The first one targets a fast-track thematic assessment of pollinators, pollination and food production (3a) (see chapter 11). The second one aims to produce three thematic assessments (3b) on land degradation and restoration, invasive alien species, and sustainable use and conservation of biodiversity and strengthening capacities/tools. The third one is related to policy support tools and methodologies for scenario analysis and modelling of biodiversity and ecosystem services (3c). The fourth one aims to identify policy support tools and methodologies regarding the diverse conceptualizations regarding biodiversity values and nature's benefits to people, including ecosystem services (3d).

The last one, *Objective 4*, aims to Communicate and evaluate Platform activities, deliverables and findings through five deliverables. The first one is a catalogue of relevant assessments (4a). The second one will produce an information and data management plan (4b), by creating a task force on knowledge and data (also responsible for deliverable 1d). The third one is a catalogue of policy support tools and methodologies (4c). The fourth one is a set of communication, outreach and engagement strategies, products and processes (4d). The last one is an evaluative deliverable and aims to review the effectiveness of guidance, procedures, methods and approaches to generate information that will be useful for future development of the Platform (4e).

Implementation dynamics

More than a year before the launch of the IPBES programme, the first steps of the Platform clearly illustrate the ways of dealing with horizontal and vertical coordination challenges (see chapter 2). The existing bodies of the Platform (the Plenary, the Bureau, the MEP and the Secretariat), all play a role in implementation of the work program, in conjunction with *ad-hoc* bodies like time-bound, task-specific expert groups and task forces, web-based arrangements and technical support units, according to the procedures below.

Work program implementation procedure

During the second Plenary, procedures for preparation of the Platform deliverables were defined (IPBES decision 2/3 in IPBES, 2013a). Three main classes of Platform assessment-related material are envisaged.

Platform reports: these include global, regional, subregional, eco-regional, thematic and methodological assessments, and synthesis reports and their summaries for policy makers. In general, Platform reports are accepted and their summaries for policy makers are approved by consensus in the Plenary. Regional and subregional reports and their summaries for policy makers are first accepted and approved by the relevant regional representatives of the Plenary and subsequently accepted and approved by the Plenary.[29]

Technical papers: these are not accepted, approved or adopted by the Plenary, but are finalized by the authors in consultation with the MEP, which serves as an editorial board.

Supporting material, including intercultural and interscientific dialogue reports: these document are not accepted, approved or adopted.

In particular, the rules define the four levels for the procedure of transforming a writing product into an official IPBES product.

Validation is, "a process by which the Multidisciplinary Expert Panel and the Bureau provide their endorsement that the processes for the preparation of Platform reports have been duly followed".

Acceptance means that the "Plenary signifies that the material has not been subjected to section-by-section or line-by-line discussion and agreement by the Plenary but nevertheless presents a comprehensive and balanced view of the subject matter".

Adoption is "a process of section-by-section endorsement at a session of the Plenary".

Approval "of the Platform's summaries for policy makers signifies that the material has been subject to detailed, line-by-line discussion and agreement by consensus at a session of the Plenary".

The IPBES procedures gave much more detail on the process of producing reports and documents, specifying the role of the Plenary, the Bureau and the MEP in relation with the experts involved, namely report co-chairs, coordinating lead authors, lead authors, contributing authors, review editors and expert reviewer authors.[30]

Assessment – the raison d'être of IPBES

Regarding the budgetary aspect, assessments represent the main provisional activity: Objective 2 will mobilize 44.5% of the estimated budget, with a predominant allocation to the global assessment (deliverable 2c), with 31%. In comparison, 25% will be spent on all of the methodological and thematic assessments planned in Objective 3.[31] This unequal distribution of the budget underlines the priority given to assessment but could lead to criticism on the lack of resources allocated to capacity building initiatives.[32]

Assessments – both thematic and geographical – are considered as a main product for IPBES. They involve the credibility, legitimacy and visibility of the new Platform. In line with adopting the global work plan, the IPBES-2 Plenary session chose to select the issue of pollination and pollinators associated with food production as a flagship for its first thematic assessment (see chapter 11). Three other thematic assessments are also included: initiation of scoping for invasive alien species and for the sustainable use of biodiversity are planned for the fourth Plenary session. The first elements of the thematic assessment on land degradation and restoration will be presented for approval at the sixth Plenary (2018).

Regarding geographical assessment, the first initiatives were centered on four regional assessments concerning the Americas, Africa, Asia and the Pacific, Europe and Central Asia. This regional approach is innovative in global environmental assessments – the option to link regional and global assessments was tested during the Millennium Ecosystem Assessment. Interestingly, some European countries like France, UK and the Netherlands have to deal with this option through a spatial distribution of their overseas territories in various regional assessments. The regional approach is also a better incentive for government cash and/or in-kind contributions to the still incomplete IPBES core budget. Conventional regional approaches do not take some specific ecosystems into account – the IPBES-3 Plenary decided to consider the option of undertaking a regional assessment for the Open Ocean region.

In methodological terms, there is a substantial challenge in managing scales between regional and global assessments of biodiversity and ecosystem services. Global assessments could be considered as the legacy of the Millennium Ecosystem Assessment and will be a key product for IPBES. The actual global assessment scoping document highlights the numerous linkages within the biodiversity regime complex and, more broadly, within the development agenda, including contributions of biodiversity and ecosystem services to the implementation of the Sustainable Development Goals, while recognizing synergies and trade-offs associated with meeting multiple goals.[33] Such global assessments are also designed as an integrated product to make effective use of most other thematic and methodological assessments.

IPBES also implements methodological assessments on two topics: (1) on scenario analysis and modelling of biodiversity and ecosystem services; and (2) on the conceptualization of values of biodiversity and nature's benefits to people.

Box 4.1 An assessment in concrete

To gain further insight into this process, we illustrate the operational structure of global assessments on multiple values of nature. These assessments are planned to mobilize three co-chairs, 80 authors and 14 review editors with the support of a technical support unit (one professional and one administrative personnel). Each of the five main chapters will include 15 core authors, including three chapter co-chairs, experts from the academic community, key stakeholder groups and ILK knowledge holders to ensure coverage of diverse worldviews. The process is supposed to be coordinated with ongoing programs such as the World Bank Waves initiative and the SEEA[34] project of the UN Statistical Office. In terms of timetable, the scoping document is to be approved in the next Plenary (IPBES-4, February 2016) and a first draft will be drawn up by late 2016 and then reviewed by experts for a second draft in May 2017. This second draft and a summary for decision makers will be submitted to governments and experts to produce a final version to be discussed in the IPBES-6 Plenary in May 2018. The estimated cost of the entire process is around US$1 M.

Extract from "IPBES – Draft Scoping Report for the methodological assessment regarding diverse conceptualization of multiple values of nature and its benefits, including biodiversity and ecosystem services – Online document for consultation" (http://www.ipbes.net/sites/default/files/downloads/For_Review_Revised_Scoping_report_for_the_assessment_of_multiple_values.pdf)

In parallel with the assessment production, IPBES manages some cross-cutting activities that aim to contribute to the various assessments, namely activities linked to the objectives of the first work plan conducted by the three task forces presented above on: (1) capacity building, (2) knowledge and data, and (3) indigenous and local knowledge (ILK) systems.

Mobilizing experts – between legitimacy and credibility

Each deliverable is under the supervision of an expert working group, supported by the Secretariat and or an *ad hoc* TSU, under the responsibility of the Bureau and the MEP, as shown by the above example on assessments on multiple values of nature. The expert mobilization dynamics were assessed by the interim MEP for the March 2013–2014 period. MEP members noted that the current nomination process did not provide adequate representation across geographic regions, gender, scientific disciplines and knowledge systems. Stakeholders suggested relaxing the 80:20 split between government and stakeholder nominations, with a shift towards up to 50% of the selections based on stakeholder nominations, and making full use of the networks that stakeholders represent. Other aspects of importance to scientific experts were definitions of the assessment acceptance, adoption and approval process, and how changes to report languages could be incorporated, and whether authors could challenge changes requested by government. A more in-depth analysis of the dynamics of expert mobilization is provided in the case study on pollinators (see chapter 11). At this stage, the challenge of achieving a balance between scientific credibility, plurality of knowledge systems and political dimensions of the process remains open. The question of conflict of interest is a good illustration of possible tensions.

IPBES has defined a conflict of interest policy since the third Plenary session (Bonn, 2015). For IPBES, a "conflict of interest" refers to any current interest of an individual that could: (1) significantly impair an individual's objectivity in carrying out his or her duties and responsabilities for the Platform, and/or (2) create an unfair advantage for any person or organization. Any request relating to a potential conflict of interest may be sent to the Platform Bureau. This policy applies to the members of the Bureau and the MEP and any other subsidiary bodies contributing to the development of deliverables, authors with responsibility for report content (including report co-chairs, coordinating lead authors and lead authors), review editors and professional staff to be hired to work in a technical support unit established by the Platform, along with the permanent professional staff of the Secretariat. Implementation of this policy relies on a conflict of interest committee consisting of three elected members from the Bureau and five regional members, one per United Nations region, with one additional member with appropriate legal expertise from and appointed by the organization hosting the Secretariat (IPBES, 2015b).

Conclusion

The development of this international science-policy interface has not been trouble free. Even though the IPCC precedent provides a very useful source of experience on this subject, establishing the Platform has been a far more complex process than that which led to the creation of the Panel. IPBES – as an intergovernmental body devoted to the science-policy interface – shares a lot of features with its often acclaimed model, the IPCC. IPBES has innovative specific aspects regarding rules and procedure, e.g. those on stakeholder engagement and participation (see chapter 8). However, this inclusive policy is under control of governmental delegations and the dominant functioning spirit remains conventionally intergovernmental, resembling the UNEP procedure.

This is in no way surprising considering the high political stakes that are shaping the whole process. North–South relations are still a sensitive topic and Rio +20, which was generally perceived as a failure, only further complicated the matter.

Some dimensions of this functioning reveal usual tensions between Global North and Global South countries. The budgetary aspects clearly show an unbalanced financial contribution, with 65% of the budget provided by two countries, i.e. Norway and Germany. The balance of influence is assumed to be effective in IPBES global governance, with a UN regional way regarding the balance of participation and with an alternate position of the Chair between a south and North candidate.

The upcoming Plenary will constitute a significant stepping stone for this young institution. Its first two finalized assessments on pollination and scenarios and models will be presented to members and observers. In a context of controversy (see chapter 11), the negotiation process for the adoption of the summaries for policy makers is likely to be heavily scrutinized. The way the Platform will apprehend this first ordeal will be crucial with regard to how its credibility, relevance and legitimacy are perceived.

Notes

1 Four Plenary sessions have been organized since the inception of IPBES: IPBES-1 in Bonn, Germany (January 2013); IPBES-2 in Antalya, Turkey (December 2013);

IPBES-3 in Bonn (January 2015); and IPBES-4 in Kuala Lumpur, Malaysia (February 2016). Regarding the establishment of IPBES (2008–2012), see chapter 3 of this book.

2 In IPBES-1 (2013), 105 countries were IPBES members, 115 in IPBES-2 and 123 in IPBES-3.

3 UNEP, 2011.

4 All citations to rule in this chapter refer to the online procedures document on the IPBES website, in particular the rules and procedures for the Plenary: http://www.ipbes.net/sites/default/files/downloads/IPBES_rules_of_procedure.pdf.

5 Group of Latin America and Caribbean Countries.

6 During the Plenary, European Union countries meet every morning to exchange information and, in some cases, coordinate their positions.

7 The rules of procedure of the Plenary states that "The Members of the Platform take decisions on matters of substance by consensus, unless otherwise provided in its rules" (IPBES1/1 Rules of Procedure for the Plenary of the Platform).

8 This option is similar to the IPCC organization with a large Bureau including three working groups: (1) the physical science basis of climate change, (2) climate change impacts, adaptation and vulnerability, and (3) mitigation of climate change.

9 Regularly, some countries highlight the underrepresentation of Antarctic and high seas regions.

10 The five UN Regions are the African Group (54 member states), the Asia-Pacific Group (53), the Eastern European Group (23), the Latin American and Caribbean Group (33), the Western European and Other Group (28).

11 R.T Watson has been Chair of the IPCC (1997–2002), Co-Chair of the Millennium Ecosystem Assessment (2000–2005) and the IIASTD Director (2002–2008).

12 Decision IPBES/1/1 Rules of Procedure for the Plenary of the Platform, rule 2.

13 Decision IPBES-2/1: Amendments to the Rules of Procedure for the Plenary with regard to rules governing the Multidisciplinary Expert Panel.

14 Since the IPBES-2 Plenary, the Secretariat is headed by Anne Larigauderie, former director of Diversitas.

15 Those three task forces contribute to Objective 1 of the work program (see next section of this chapter).

16 All the data come from IPBES 2015b: 59–72.

17 In January 2015, IPBES had received in-kind offers from 12 governments and 24 organizations.

18 In comparison, the Millennium Ecosystem Assessment yearly budget was estimated at US$3,200,000.

19 Decision IPBES/1/4 IPBES administrative and institutional arrangements.

20 Decision IPBES-2/8: Collaborative partnership arrangement to establish an institutional link between the Plenary and UNEP, UNESCO, FAO and UNDP.

21 UNEP/CBD/COP/DEC/X/11, Science-policy interface on biodiversity, ecosystem services and human wellbeing and consideration of the outcome of the intergovernmental meetings, Nagoya, October 18–29, 2010.

22 UNEP/CBD/SBSTTA/19/9, Work of the Subsidiary Body on Scientific, Technical and Technological Advice in the Light of the 2014–2018 Work Program of the Intergovernmental Science-Policy Platform for Biodiversity and Ecosystem Services and Relationship with the Subsidiary Body on Implementation, November 2–5, 2015.

23 UNEP/CBD/COP/DEC/XII/25, Intergovernmental Science-Policy Platform on Biodiversity and Ecosystem Services, October 6–17, 2014.

24 UNEP/CBD/COP/DEC/XII/25, *op. cit.*, §.1.

25 AC28 Doc. 6.1, Intergovernmental Science-Policy Platform on Biodiversity and Ecosystem Services (IPBES) (Decision 16.15), August 30 to September 3, 2015.

26 Between two plenaries, the members of the Bureau, the MEP and other experts involved in the various task forces or working groups participate in the intersessional meetings.

27 This option was decided in Busan meeting (2010): other existing options discussed at this meeting envisage to manage requests through CBD only, or the six biodiversity-related conventions or multilateral environmental agreements related to biodiversity and ecosystem services and United Nations agencies.
28 All those deliverables are linked to the Aichi Biodiversity Target (18 and 19).
29 "Approval" means that the material has been subject to detailed, line-by-line discussion and agreement by consensus at a Plenary session. "Adoption" is a process of section-by-section endorsement and "Acceptance" at a Plenary session means that the material has not been subjected to line-by-line discussion and agreement, but nevertheless presents a comprehensive and balanced view of the subject matter.
30 See IPBES 2013b: 16–38.
31 Based on the estimated cost of the work program in IPBES 2013b (IPBES/2/17).
32 The estimated budget for capacity building activities (deliverable 1a and 1b) is around 14% of the estimated cost of the work program (IPBES 2013b: 64).
33 Draft document for discussion on the scoping report for the IPBES Global Assessments of Biodiversity and Ecosystem Services (Deliverable 2c) (2015).
34 System of Environmental-Economic Accounting.

References

Boyle, A., Chinkin, C. 2007. *The Making of International Law*. Oxford: Oxford University Press.

Brunnee, J. 2002. COPing with Consent: Law-Making under Multilateral Environmental Agreements. *Leiden Journal of International Law*, vol. 15, 1–52.

CBD 2010. Guide for working group chairs at COP and SBSTTA meetings. Secretariat of the Convention on Biological Diversity.

ENB 2012. The second session of the plenary meeting on an Intergovernmental Platform on Biodiversity and Ecosystem Services (IPBES), 15–21 April 2012. Earth Negotiations Bulletin, International Institute for Sustainable Development (IISD).

ENB 2013a. Summary of the first plenary meeting of the Intergovernmental Platform on Biodiversity and Ecosystem Services, Bonn, 21–26 January 2013. Earth Negotiations Bulletin, vol. 31, n°1, International Institute for Sustainable Development (IISD).

ENB 2013b. Summary of the second plenary meeting of the Intergovernmental Platform on Biodiversity and Ecosystem Services, Antalya, 9–14 December 2013. *Earth Negotiations Bulletin*, vol. 31, n°7, International Institute for Sustainable Development (IISD).

ENB 2015. Third Session of the Plenary meeting of the Intergovernmental Platform on Biodiversity and Ecosystem Services, 12–17 January 2015. Earth Negotiations Bulletin, International Institute for Sustainable Development (IISD).

IPBES 2013a. Report of the first session of the Plenary of the Intergovernmental Science-Policy Platform on Biodiversity and Ecosystem Services (IPBES/1/12), Bonn, 21–26 January.

IPBES 2013b. Report of the second session of the Plenary of the Intergovernmental Science-Policy Platform on Biodiversity and Ecosystem Services (IPBES/2/17). Antalya, Turkey: United Nations Environment Programme (UNEP).

IPBES 2013c. Rules of Procedure for the Plenary of the Platform (IPBES). United Nations Environment Programme (UNEP).

IPBES 2015a. Guidance document on the nomination and selection process for members of the Multidisciplinary Expert Panel and lessons learned from the experience of the interim Panel (IPBES/3/INF/16). UNEP, UNESCO, FAO, UNDP.

IPBES 2015b. Report of the Plenary of the Intergovernmental Science-Policy Platform on Biodiversity and Ecosystem Services on the work of its third session (IPBES/3/18). Bonn, Germany: United Nations Environment Programme (UNEP).

Lachs, M., 1972. Some Reflections on Substance and Form in International Law. *Transnational Law in a Changing Society. Essays in Honour of Philip C. Jessup*. New York: Cambridge University Press, p. 100.

Montana, J., Borie, M. 2015. IPBES and Biodiversity Expertise: Regional, Gender, and Disciplinary Balance in the Composition of the Interim and 2015 Multidisciplinary Expert Panel. *Conservation Letters*, 1–5.

Morin, J.-F., Louafi, S., Orsini, A. & Oubenal, M. 2015. Boundary Organizations in Regime Complexes. A Social Network Assessment of IPBES. *International Studies Association*. New Orleans.

Thibierge, C., 2009, La *force normative. Naissance d'un concept*. Paris: LGDJ.

UNEP 1988. Rules of procedure of the Governing Council (UNEP/GC/3/Rev.3, 22p).

UNEP 2008. Governance structure and secretariat functions for an intergovernmental science-policy platform on biodiversity and ecosystem services (UNEP/IPBES/1/4). Kuala Lumpur, 10–12 November 2008.

UNEP 2010. Report of the third ad hoc intergovernmental and multi-stakeholder meeting on an intergovernmental science-policy platform on biodiversity and ecosystem services (UNEP/IPBES/3/3). Busan, 7–11 June 2010, Annex I.

UNEP 2011a. Report of the Governing Council of the United Nations Environment Programme on its eleventh special session (A/RES/65/162). 15 March 2011.

UNEP 2011b. Decisions adopted by the Governing Council/Global Ministerial Environment Forum at its 26th session (UNEP/GC.26/19), p.9. Decision 26/4: Intergovernmental science-policy platform on biodiversity and ecosystem services.

UNEP 2011c. Legal opinion of the Secretariat concerning certain legal issues relating to the establishment and operationalization of the Platform (UNEP/IPBES.MI/INF/9). Nairobi, 3–7 October 2011, §.4.

UNEP 2012a. Functions, operating principles and institutional arrangements of the Intergovernmental Science-Policy Platform on Biodiversity and Ecosystem Services (IPBES). United Nations Environment Programme (UNEP).

UNEP 2012b. Report of the second session of the plenary meeting to determine modalities and institutional arrangements for an intergovernmental science-policy platform on biodiversity and ecosystem services (UNEP/IPBES.MI/2/9). Panama City: United Nations Environment Programme (UNEP).

UNEP 2012c. Resolution on the Intergovernmental Science-Policy Platform on Biodiversity and Ecosystem Services (IPBES). United Nations Environment Programme (UNEP).

UNEP 2013. Report of the first session of the Plenary of the Intergovernmental Science-Policy Platform on Biodiversity and Ecosystem Services (IPBES/1/12). Bonn, Germany: United Nations Environment Programme (UNEP).

5   From climate to biodiversity

Procedural transcriptions and innovations within IPBES in the light of IPCC practices

Guillaume Futhazar[1]

Describing IPBES[2] as simply an 'IPCC[3] for biodiversity' would be overly simplistic. Indeed, these two institutions have major differences that have been underlined on several occasions.[4] These differences are the result of fundamental structuring aspects: the respective themes of these institutions, the institutional context surrounding their establishment and their mandate.

Concerning their themes, it is no longer necessary nowadays to engage in a lengthy demonstration to illustrate that climate and biodiversity[5] are facing major crises that call for swift and global responses. However, although climate change is the prime example of a global crisis, biodiversity constitutes an element that is intrinsically linked to territories as each region has to deal with specific yet connected threats. Moreover, biodiversity is not uniformly distributed among territories. Some countries, known as 'mega-diverse', have a higher degree of biological diversity and a vast majority of these are developing countries. As such, we can see through this notion the resurgence of the South/North divide in environmental debates. This strong link between biodiversity and territories also allows for a salient influence of the concept of permanent sovereignty over natural resources. Debates and negotiations for the establishment of the Intergovernmental Platform on Biodiversity and Ecosystem Services (IPBES) and within its fora are clearly heavily influenced by these eminently political and legal considerations.

IPBES and the Intergovernmental Panel on Climate Change (IPCC) were negotiated and established in very different institutional contexts. IPCC was founded prior to the adoption of the United Nation Framework Convention for Climate Change (UNFCCC)[6] and it even took part in discussions that led to the Convention's adoption,[7] whereas IPBES has had to fit into in a very crowded institutional environment. Major conventions on biodiversity and ecosystems have existed for decades and constitute a regime complex, which is understood as being an accumulation of overlapping regimes with no hierarchy[8] and various types of interaction.[9] Alongside these conventions, several UN institutions are also actively working on this broad theme, most notably UNEP, UNDP, FAO and UNESCO[10] (which are among the main IPBES partners). The establishment of IPBES therefore raises the necessary question of how it will interact with numerous actors of the system.

Finally, the IPBES mandate is not a simple reiteration of what IPCC does. Both IPCC and IPBES seek to periodically assess the state of knowledge for their respective themes, but IPBES goes further by providing innovative tools and advice for its members.[11] The Platform, based on its work on the different assessments, will produce a list of research priorities. It will also work on strengthening capacity building and identifying tools and methodologies for decision makers in order to achieve effective biodiversity governance. However, like IPCC, IPBES will not generate new knowledge on its own.[12]

These several major differences highlight the fact that we must abandon the idea that IPBES is only a duplication of the IPCC experience on a different topic. Still, it would be irrelevant to completely disconnect these two institutions as they are obviously interlinked. The intervention of the IPCC ex-Chair Rajendra Pachauri during the 3rd IPBES Plenary is an illustration, among others, of the links between the two bodies. Climate and biodiversity are also connected elements and the biodiversity crisis will not be halted if the climate crisis is not dealt with.[13] This situation has been underlined by the Millennium Ecosystem Assessment which demonstrated that climate change is one of the drivers of biodiversity loss and its impact will increase in the upcoming years.[14] Biodiversity is also a tool in the mitigation of climate change as healthy ecosystems can absorb carbon and reduce its accumulation in the atmosphere.[15] Moreover, negotiators involved in the discussion for the establishment of IPBES have been clearly influenced by IPCC's experience despite stressing the differences between the two initiatives. Overall, we can see a tendency within IPBES to simultaneously differentiate itself from its model while also reaffirming its evident ties.

The purpose of the present contribution is to study how this complex situation has impacted the IPBES rules of procedure by comparing them with those that prevail within the IPCC framework. However, prior to this analysis, it is essential to establish a precise definition of the term 'procedure' in the context of this study.

What is a procedure?

Most legal research is focused on a specific definition of the concept of procedure, based on its contentious aspect. Procedures are usually understood as being the rules shaping the trial process. However, in the present chapter, our understanding of the term is broader and closer to the general definition – basically the steps to follow in order to accomplish a certain task or achieve a specific result.[16] As such, this definition concerns the functioning of institutions and covers rules governing their internal functioning and shaping their activities. However, this definition does not provide any indication regarding the legal consequences of these procedures.

In the context of IPBES, a breach of the procedures would not have any legal consequences. Neither the rights nor the obligations of member states would be impacted by a procedural breach and no third party could be seized with this matter in order to pronounce sanctions. Yet, this absence of legal consequence does not imply that the application of these procedural rules is optional. The Platform and any other relevant actors playing a part in the process are bound by them and their application ensures that the activities and productions of the Platform are perceived as being legitimate. Indeed, rules of procedure are the product of state consensus with states being the only parties entitled to adopt them within the Platform Plenary.[17] As a consequence of the consensus requirement, the Platform Plenaries generated lengthy negotiations in order to lay down the procedural aspect of this new institution. The rules of procedure are therefore the expression of the member states' expectations on how the Platform must work. Any production or activity that would overstep the procedural framework would consequently be rejected by the members. In short, compliance with the agreed procedures is the necessary condition for the legitimacy of the IPBES process towards its members, with legitimacy being understood as what is perceived as being fair, equitable and politically acceptable.[18]

The procedures not only ensure the legitimacy of the activities and deliverables of the Platform towards the member states, they also do so towards stakeholders. Indeed, procedures as well as all of the IPBES institutional characteristics (mandate, structure, and work programme) tend to fulfil three main criteria:[19] credibility,[20] relevance[21] and legitimacy. Therefore, this chapter also aims to determine whether the procedures of this young institution satisfy these principles.

In discussing the diffusion and adaptation of procedure from IPCC to IPBES, this chapter echoes the work of Alexander Ovodenko and Robert Keohane, who established a theoretical framework for institutional diffusion, which is understood as being, 'the adoption in new or reformed institutions of institutional features already operating in other institutions, national, international or transnational'.[22] The authors identify two factors that must be taken into account when studying diffusion cases. The first factor is functional: an institutional characteristic, such as a rule of procedure, will be diffused if the two institutions face the same problem structure, have similar involved actors, and if the institutional characteristic being diffused is perceived as being successful.[23] However, this first factor is insufficient on its own to fully understand institutional diffusion and we must also take into account the political factor, which encompasses the different ideas, powers and interests coexisting within a system.[24] The two different sections of this chapter reflect these two factors. The first section illustrates the obvious influence IPCC has had on the establishment of the IPBES rules of procedure for functional reasons (1), and the second section highlights the procedural innovations of the Platform and attempts to explain them in the light of the Platform characteristics (2).

Overall this chapter hopes to add to the debate concerning the linkages between IPCC and IPBES[25] by using, as a starting point, elements that are seldom the focus of detailed analysis: rules of procedure.

The influence of IPCC in the shaping of IPBES

During the lengthy discussions that led to the establishment of IPBES, one main argument was put forward on several occasions, i.e. that the Platform is not a simple reiteration of IPCC. During the Putrajaya meeting in 2008, the participants insisted on this aspect while also recognizing that IPCC could serve as a model in the debates on how to design the new institution.[26] This call for differentiation yet inspiration recurred at each meeting that addressed the creation of the Platform and even during the IPBES Plenary. For instance, during the Busan meeting in 2010, the fact that an information document on IPCC was distributed to the attendees is particularly significant as it illustrates the status of IPCC's experience in negotiators' minds.[27] This situation evidently led to several procedural transplantations from IPCC to IPBES: firstly, the IPBES report drafting process and secondly the conflict of interest policy.

The IPBES report drafting process

The IPCC report drafting process is the result of a slow process that began following its creation in the late 1980s.[28] Several political factors led IPCC to further elaborate procedures that would ensure the highest possible credibility for its reports. As such, the transplantation of these procedures in the context of IPBES is clearly an asset for the Platform, which is able to benefit from the vast experience of IPCC in this field.

A close look at the procedures of both IPCC and IPBES shows how similar they are, even to the point where some dispositions appears to be clear 'copy/paste' of IPCC rules. For instance, the procedure for the incorporation of reviews, which aims to ensure greater credibility and scientific soundness, is a practically identical step for both organizations (see Table 5.1).

This comparison could be made for several aspects of the drafting procedures, thus showing how the IPCC precedent has had a huge influence in the conceptualization of the Platform procedures. Obviously, aspects calling for adaption because of the Platform's specificity have been modified accordingly, but nevertheless there is clearly a high degree of similarity in the way both institutions produce their reports (as illustrated in Figure 5.1).

This transplantation is a token of credibility for the Platform. IPCC procedures have undergone the test of time and controversies, and they guarantee the current status of the Panel as a major scientific actor in the field of climate change. Hence, it can be hoped that this credibility will be as strong for the Platform.

Prevention and management of conflicts of interest

Dealing with conflicts of interest is of crucial importance in preventing a legitimacy and credibility crisis. Any report accused of bearing skewed information because of a conflict would be immediately rejected by both

Table 5.1 Comparison of procedures for the incorporation of comments in the drafting of IPBES and IPCC reports

IPCC[29]	IPBES[30]
4.3.4 Review	3.6.4 Review
Three principles governing the review should be borne in mind. First, *the best possible scientific and technical advice* should be included so that the IPCC Reports represent the latest scientific, technical and socio-economic findings and are as comprehensive as possible. Secondly, a wide circulation process, *ensuring representation of independent experts (i.e. experts not involved in the preparation of that particular chapter) from developing and developed countries* and countries with economies in transition should aim to involve as many experts as possible in the IPCC process. Thirdly, *the review process should be objective, open and transparent.*	Three principles govern the review process: first, the Platform reports should represent *the best possible scientific, technical and socioeconomic advice* and be as balanced and comprehensive as possible. Second, as many experts as possible should be involved in the review process, *ensuring representation of independent experts (i.e., experts not involved in the preparation of the chapter they are to review) from all countries.* Third, *the review process should be balanced, open and transparent* and record the response to each review comment.

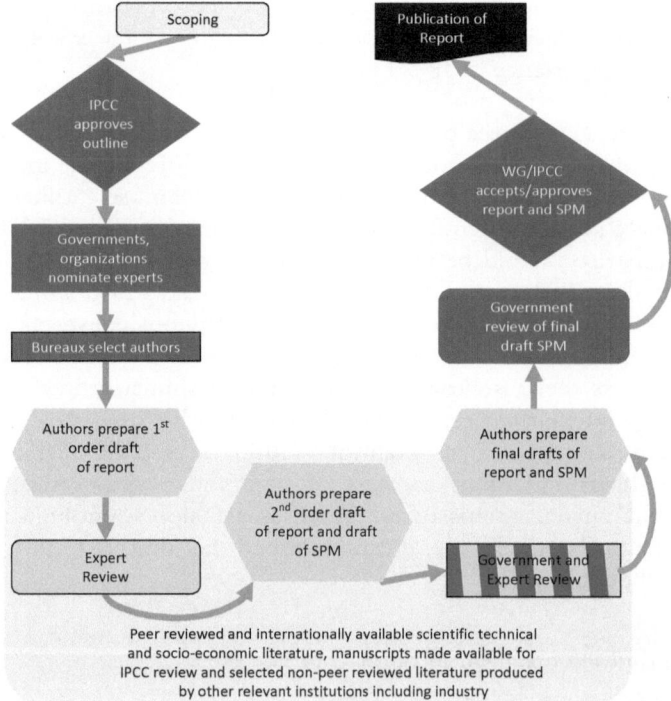

Figure 5.1 Writing and review procedures for IPCC
www.ipcc.ch; simplified version by the author

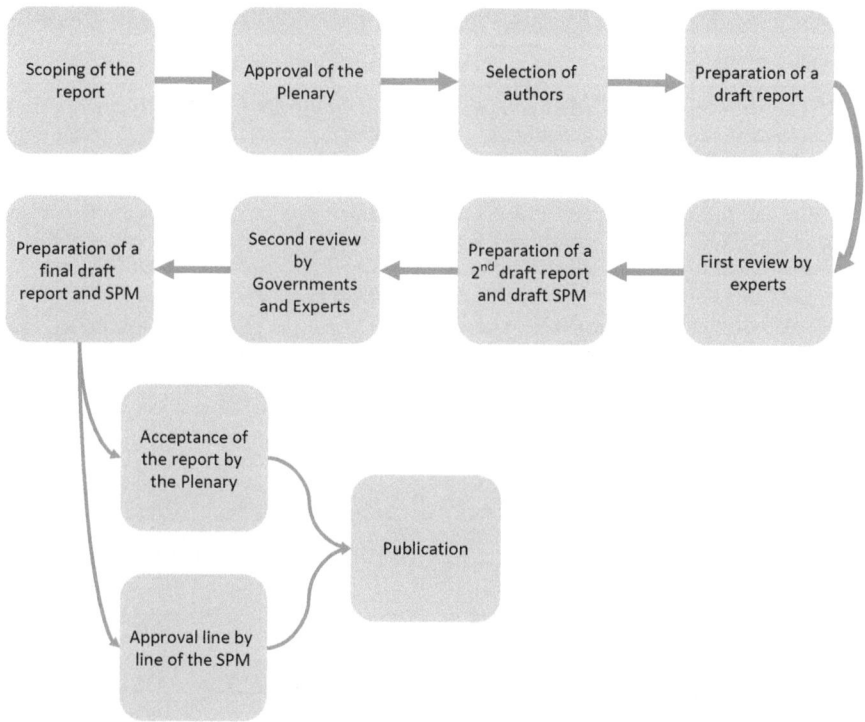

Figure 5.2 Writing and review procedures for IPBES, simplified version by the author

members and stakeholders. The existence of the conflict would be perceived as a breach of equity and justice and all related content of the report would be dismissed as being 'false' or at least not trustworthy.

IPCC has gone through a severe crisis in recent years. The mistakes that were pointed out in its report and the strong opposition of opponents led the Panel to reflect on its procedures. The mandate given to the Inter Academic Council to produce an independent review of the Panel led to numerous changes,[31] including the adoption of procedures for the prevention and management of conflicts of interest in 2011.[32]

These procedures were later incorporated in the IPBES framework during its third Plenary. This incorporation came at a crucial time when the Platform was already confronted with its first controversies. It was accused of promoting the interest of industry by allowing scientists involved in the pesticide production sector to be in charge of the drafting of chapters in the upcoming report on pollination.[33] The IPBES Secretariat responded to this accusation by stressing that the Platform procedures were suited to manage any situation that could lead to a conflict of interest.[34]

By analysing both procedures, we can once again see that the two institutions adopted similar approaches (see Table 5.2). Several definitions are

Table 5.2 Partial comparison of the IPBES and IPCC conflict of interest policies

IPCC[36]	IPBES[37]
§. 2: The role of the IPCC demands *that it pay special attention to issues of independence and bias in order to maintain the integrity of, and public confidence in, its products and processes. It is essential that the work of the IPCC is not compromised by any conflict of interest for those who execute it.*	§. 2: The role of the Platform requires *that it pay special attention to issues of independence and bias in order to maintain the integrity of, and public confidence in, its products and processes. It is essential that the work of the Platform is not compromised by any conflict of interest for those who execute it.*
§. 11: A 'conflict of interest' refers to any current professional, financial or other interest which could: i) significantly impair the individual's objectivity in carrying out his or her duties and responsibilities for the IPCC, or ii) create an unfair advantage for any person or organization. *For the purposes of this policy, circumstances that could lead a reasonable person to question an individual's objectivity, or whether an unfair advantage has been created, constitute a potential conflict of interest.* These potential conflicts are subject to disclosure.	§. 11: *For the purposes of this policy, any circumstances that could lead a reasonable person to question either an individual's objectivity, or whether an unfair advantage has been created, constitute a potential conflict of interest.*

taken from the IPCC framework and the overall mechanism is identical (establishment of a committee, requirement that experts fill in a form). However, caution is needed when claiming that these procedures are sufficient to ensure complete credibility. Indeed, the screening process of experts is far from being strict. The form the experts are required to fill in is very simple and requests any information about, for instance, any previous professional ties that the expert may have had.[35] Moreover, the fact that these forms are kept secret by the Platform can be seen as a breach of transparency which might have an impact on the credibility of the process. Yet, total access to the form would be a breach of the experts' privacy. Finding balance on this issue is rather complicated and only time will tell if this rule of procedure is sufficient for efficient management of interests within the Platform. The fact remains that only IPBES is able to determine whether or not a conflict of interest is real or not, which might fuel dissent amongst observers.

The list of similarities between the two institutions could go on at length and this first section makes no claim to be exhaustive. The purpose of this short list is to simply illustrate the degree of influence the Panel has on the day-to-day work of the Platform.

If the examples used underline the relevance of the functional factor of institutional diffusion, the next section illustrates the influence of external factors in the shaping of the procedural identity of the Platform.

Distinction and innovation – the creation of a distinct procedural identity

As previously stated, political factors are to be taken into account when studying institutional diffusion. The introduction of this chapter recalled the marked differences in the context of each institution. These contexts bear different interests, powers and ideas and have a significant influence on how the IPBES rules of the procedure are constructed. The present section aims at discussing how this context led to the adoption of procedural innovations but was also the cause of unwelcomed modifications of rules transplanted from IPCC to IPBES.

Ambitious procedural innovations

The most salient innovations concern the way the Platform interacts with its members and stakeholders. In order to promote legitimacy and relevance, IPBES has adopted procedures guaranteeing collaboration and openness. In this respect, the two most remarkable procedural aspects are the working programme drafting process on the one hand and the formal institutional linkages with other partners on the other.

Elaboration of the work programme

The way the Platform work programme is established illustrates how the numerous expectations of the actors involved in the creation of IPBES are taken into account.

In the IPCC framework, the way the work programme is set remains constant from one period to another. Each evaluation cycle will discuss three aspects of climate change: 1) the scientific evidence, 2) the impact, adaptation and vulnerability, and 3) mitigation. However, this consistency does not imply strict similarity between each Panel report. IPCC rules of procedure state that the exact outline of each evaluation cycle will be determined by experts from governments and observer organizations and Bureau members.[38] Once this outline is agreed by the Panel Plenary, the evaluation process is applied as previously described. Based on this rationale, IPCC has published five reports since its creation (roughly one every 5–6 years) and has contributed to strengthening scientific consensus on the role of mankind in climate change, while also highlighting available solutions to reduce or cope with a phenomenon that is a dire threat to all societies.[39]

However, in the IPBES context, the work programme elaboration process allows for variation between evaluation cycles. Indeed, the exact content of the work programme is determined by the Multidisciplinary Expert Panel (MEP) which relies on requests, suggestions and inputs from members, observers and stakeholders.[40] All of these requests are then synthesized by the MEP in order to produce a first draft of the work programme that is subsequently discussed by members during the Plenaries and eventually adopted as the definitive work programme. In this synthesis process, priority is given to the requests of states while stakeholder requests, i.e. from international organizations, multilateral environmental agreements (MEAs) or NGOs, are considered as suggestions or inputs. This balance in favour of states is a logical repercussion of the intergovernmental aspect of the Platform. The outcome of this process resulted in the adoption of the first Platform work programme during the second Platform Plenary in 2014.[41] For instance, this supplementary step in drawing up a work programme, which does not exist in the IPCC framework, contributed to reaching an agreement on the regional scope of evaluations that the Platform intends to produce in the upcoming years, or even on the elaboration of a report on land degradation and restauration.

The diverse requests received by the Platform illustrate how the current biodiversity and ecosystem crisis has several translations within different levels and sectors (Table 5.3).

States often have to deal with specific issues that can considerably vary between regions. For instance, France formulated a request on the protection and sustainable use of marine ecosystems, whereas China requested an evaluation on biodiversity and ecosystems in post-disaster areas. The Convention on Migratory Species requested an assessment of the ecological function of migratory species, and other multilateral environmental agreements also issued a request to the Platform to work on their respective thematic foci. Stakeholders had various suggestions, from the establishment of a long-term framework for biodiversity data provision and use (request by the Global Biodiversity Information Facility) to actions to strengthen the use of scenarios and models in assessments (suggestion by the International Council for Science).

This procedure highlights the Platform's efforts to take the wide range of expectations from its members and stakeholders into account. By doing so, the work programme is likely to be perceived as highly relevant. Moreover, states are not the only actors who expect the Platform to have useful outcomes. Biodiversity-related conventions, international organizations, NGOs and research institutions have the same expectations. Even though their inputs are not a priority when drafting the work programme, they are

Table 5.3 Overview of requests, suggestions and inputs received by the Platform for elaboration of the first work programme[42]

	Number of actors	Number of requests, suggestions and inputs
States	10	36
Multilateral environmental agreements	4	13
Stakeholders	10	30

nevertheless taken into account by MEP. All of these procedural elements, if applied correctly, will be a strong asset for the relevance of the Platform. Hopefully the Platform will receive more requests, suggestions and inputs when the time comes to draw up a second work programme. Indeed, for an intergovernmental institution with over a hundred member states, it is unfortunate that only 10 states have made requests. Although the work programme was discussed amongst the members and is therefore the result of consensus, higher member involvement in this first phase would increase the legitimacy and relevance of the next work programmes.

As illustrated by the work programme elaboration process, IPBES has links with numerous actors working on biodiversity and ecosystems governance. This important feature led to the adoption of several decisions to establish the Platform's external relations management procedures.

Formal ties with other actors of biodiversity and ecosystem governance

The decisions adopted in 2015 illustrate the extent to which the implementation of the work programme is expected to be a collaborative effort. The place allocated for stakeholders within the Platform is a salient illustration of this aspect.[43] Indeed, member states invite stakeholders to organize themselves within an open-ended network in order to support the work of the Platform by helping with the nomination of experts or liaising with hard to reach stakeholders. This call for greater involvement of the protean community of stakeholders is both praiseworthy and necessary as it tends to ensure credibility and legitimacy.[44] Even if this aspect cannot be perceived as procedural according to our understanding of the term, one aspect of its implementation is linked to the procedures for the establishment of strategic partnerships. Indeed, the decision on the stakeholder engagement strategy calls for the adoption of a strategic partnership during the next Platform Plenary. In the context of our analysis, the rules of procedure adopted on this topic highlight how crucial the status of other actors of biodiversity and ecosystem governance is for IPBES.[45]

Strategic partnerships have several purposes, one of which is the creation of technical support units (TSUs). Like IPCC, IPBES relies on these TSUs in charge of coordinating activities linked to the production of its deliverables. IPBES has already established such units with the support of UNESCO on the theme of indigenous and local knowledge.[46] This practice is yet another example of institutional transplantation from IPCC. However, IPBES is innovative on the administrative aspects of these TSUs,[47] which are established through very specific agreements that clearly state both the precise mandate of the unit and the responsibilities of the Platform and the unit. This aspect should allow for a more effective TSU governance in the coming years.

However, strategic partnerships are not only meant to establish TSUs. The decision on strategic partnerships also calls for formal collaboration with the other major multilateral environmental agreements in the field of biodiversity and ecosystems. Partnerships should facilitate mainstreaming of the Platform and MEA activities and avoid redundancy. This is also a way for the Platform to ensure the relevance of its activities towards MEAs.[48] On the other hand, in the IPCC context, the Panel has no formal framework of collaboration with UNFCCC. This issue has been addressed several times within IPCC and has yet to be definitively resolved.[49]

In its application of the decision on strategic partnership, IPBES relies on several types of legal instruments in order to formalize its relations with other institutions. Memorandums are frequently used. These pseudo-contracts remain a mystery in the field of legal studies.[50] Written in the form of a contract, they do not possess any legal strength and only allow for the formalization of cooperation principles between several actors. Yet, their contractual aspect often includes very precise articles on interpretation, entry into force, settlement of disputes and even denunciation.

The use of such instruments, which elude classical legal conceptions yet are so close to them, adds a supplementary dimension to the way IPBES interacts with its institutional environment as compared to IPCC. This is a very interesting example of how a procedural requirement translates into the use of quasi-legal or para-legal instruments. It will be important to check on a practical level how far this aspect will influence future development of the Platform.

In many respects, IPBES is innovative when it comes to its interaction with other actors of the regime. These innovations can be understood as a consequence of the very rich institutional landscape within which the Platform is established. However, besides those positive innovations, some worrying modifications have been made in the transplanted IPCC procedures that may undermine the credibility of the Platform.

An unfortunate change in the expert selection process

During the second Platform Plenary, criteria for the selection of experts were at the centre of a heated debate.[51] This question has a very important strategic aspect that led to an opposition between two rationales. On the one hand, some states, including China and Argentina, wanted total control over the nomination of experts, thus keeping any expert nominated by relevant institutions that did not have a formal endorsement by the state out of the process. This condition would put IPBES in a very different position than IPCC, where experts are nominated by both member states and relevant institutions.[52] On the other hand, European Union states, with support from the attending observers, wanted a process similar to that of IPCC in order to nominate experts most able to ensure satisfactory implementation of the work programme.

The push towards state control over the nomination of experts was one of the crucial points of the second Plenary. The outcome of negotiations on this point gave rise to a convoluted solution whose application is not straightforward.

MEP members, nominated by member states according to a geographical balance,[53] have the responsibility of selecting experts in charge of the drafting of each work programme report. They do so by relying on a list of experts nominated by both governments and relevant institutions. However, the selection has to respect a very precise balance that calls for a quota of 80% of government nominated experts and 20% of experts from relevant institutions.[54] Following this rule of procedure, if a report requires 100 experts to be drafted, 80 of them have to be government nominated.

This procedural requirement illustrates the extent of the stakes associated with biodiversity and ecosystems, most notably state sovereignty and the South/North divide. For instance, states in favour of a higher control over the expert selection process claimed that it was of the highest importance that evaluations concerning their biodiversity and ecosystems be led by their own experts. This situation clearly demonstrates how difficult it is to focus on scientific aspects alone in international negotiations even when the rights and obligations of states are not directly concerned. In summary, this is a case of diffusion from IPCC to IPBES but with a modified outcome due to the ideas and interests of specific states in the Platform context.

This procedural aspect leads to obvious practical issues. Its strict application makes it hard to achieve the required number of selected experts. The IPBES Secretariat highlighted this problem in its report on implementation of the work programme and suggested to member states that they should reconsider this procedural rule.[55] For most European states, this 80/20 requirement does not constitute a severe obstacle for the nomination and selection of experts as they often closely collaborate with their own national institutions when nominating experts. However, in other regions, most notably Eastern Europe, this rule of procedure hinders the selection process. For instance, when a group of states has not nominated a sufficient number of experts to reach the required 80%, it is necessary to call for additional nominations within this group. These 'default' nominations can be a drawback for the credibility of the Platform as the involved experts might not be as qualified or motivated as the stakeholders and observers hope them to be. This can also be a threat to the quality of the deliverables in which those experts were involved.

Moreover, considering the fact that the nomination of experts is already influenced by the heavy *pro bono* workload that the involvement in the IPBES process represents,[56] one can only be sceptical towards this procedural aspect that may even further limit the involvement of qualified and willing experts. As such, this rule of procedure appears to be in contradiction with state expectations for the Platform. Indeed, during the second and third Plenaries, on numerous occasions, states pointed out that the work programme may be too ambitious and too heavy on IPBES. Yet, even though they were fully aware of this aspect, several of their decisions on the budget or the procedures ended up being more of an impediment than a support for the Platform. The fact that they did not wish to change the 80/20 requirement for the selection of experts illustrates this ambiguous discourse by Platform members who ask so much of it while reducing its capacity. We will have to wait until the end of the first work programme to determine whether or not this analysis is too pessimistic, but from a theoretical standpoint, this procedural adaptation is a cause for concern.

Conclusion

This chapter has described a situation that is simply a logical reflection of the external elements which have shaped the establishment of the Platform. To overlook the IPCC case would have meant overlooking the wealth of experience amassed over time and by weathering the storms of controversy. How could one not acknowledge this institution when the initial idea for the creation of the Platform was to form an 'IPCC for biodiversity'? Not taking these precious lessons into account would have been a serious mistake. The current IPBES rules of procedure are, as such, very similar to those of IPCC. This does not mean however that IPBES is bound to simply be a copy of IPCC.

IPBES is innovative in several ways. Although this chapter has focused on its procedure, the Platform has also adopted some very interesting governance tools, such as the 'conceptual framework'.[57] This framework illustrates how IPBES differentiates itself from IPCC not only by its procedure but also by the concepts upon which its work relies. Whereas IPCC is currently working on how to integrate traditional and local knowledge in its work, this theme was integrated very early in the IPBES framework via the conceptual framework. Even though this framework is not within the scope of our analysis, illustrating how distinct IPBES is from IPCC would have been incomplete without referring to this aspect.

Some observers consider that currently IPBES activities are too similar to those of IPCC and that it should seek to differentiate itself more.[58] This opinion is based on the fact that half of the IPBES budget is allocated to the evaluation function of the Platform, whereas the other half is divided among its other functions. However, considering the rules of procedure of the Platform and its context, it could be expected that in the years to come IPBES will develop in a very different way from IPCC. Now that the Platform has 'started working', it will forge its own identity. Moreover, the two institutions actually have started exchanging on several topics and the purpose of these exchanges is not only for IPBES to learn from IPCC. During its recent Plenary in Nairobi, IPCC took the practice of IPBES concerning its TSUs into account and suggested that its members should adopt new procedures based on this model.[59] Even though this proposal was turned down due to the lack of perspective on the long term application of these rules of procedures,[60] it goes to show that future relations between these two institutions will surely be based on mutual learning. This perspective of coevolution is a good sign as it will participate in elevating the work standards and also contribute to ensuring the highest credibility, legitimacy and independence.

Yet, the Platform is not above criticism. The criteria adopted for the selection of experts, its relatively low budget and heavy workload may hinder implementation of its work programme. As such, before putting forward hypotheses on the future influence of the Platform on IPCC, it is essential to first follow its initial process and assess its outcome. The adoption of its first reports in February 2016 will be a first trial and will pave the way for its future development.

Although this chapter has analysed the influence of various external elements, it cannot offer any definitive proposals on the future of the Platform. It would be safe to assume that it will forge its own institutional identity since its mandate and procedures are obviously distinct from its model. However, considering the very ambitious work programme set by its members and the numerous obstacles it has to overcome, the prospects give cause for concern. The next meeting in Kuala Lumpur may provide answers to all of these interrogations.

Notes

1 PhD candidate on contract for Labex OT-MED (ANR-11-LABX-0061) under the supervision of CERIC (CERIC–CNRS UMR 7318, Aix Marseille University, France). This chapter is a contribution to the CIRCULEX project (ANR-12-GLOB-0001–03 CIRCULEX), funded by the French National Research Agency, which is devoted to studying the circulation of international environmental governance norms and actors.

2 Intergovernmental Platform on Biodiversity and Ecosystem Services.

3 Inergovernmental Panel on Climate Change.

4 See Brooks, T.M., Lamoreux, J.F., Soberon, J., 'IPBES≠IPCC', *Trends in Ecology and Evolution*, vol. 29, n° 10, 2014, pp. 543–545; Maljean-Dubois S. 'La Plateforme Intergouvernementale Scientifique et Politique sur la Biodiversité et les Services Ecosystémiques (IPBES)', *Journal International de Bioéthique*, vol. 35, n° 1, 2014.

5 In the context of this chapter, we understand biodiversity as being 'the variability among living organisms from all sources' (article 2 of the Convention on Biological Diversity).

6 United Nations Framework Convention on Climate Change, New York, 21 March 1992, entered into force on 9 June 1994, *U.N.T.S* vol. 1771, p. 107.

7 Agrawala, S., 'Structural and Process History of the IPCC', *Climate Change*, vol. 39, 1998, pp. 621–642.

8 Rausitalia, K., Victor, D., 'The Regime Complex for Plant Genetic Resources', *International Organization*, vol. 58, n°2, 2004, pp. 277–309.

9 Orsini, A., Morin, J.F., Young, O., 'Regime Complexes: A Buzz, a Boom, or a Boost for Global Governance?', *Global Governance*, vol. 19, 2013, pp. 27–39.

10 United Nations Environment Programme; United Nations Development Programme; Food and Agriculture Organization of the United Nations; United Nations Educational, Scientific and Cultural Organization.

11 The detailed mandate of the IPBES is available online: http://www.ipbes.net/poli cies-and-procedures (last accessed: 30/06/2015).

12 Functions, operating principles and institutional arrangements of the Intergovernmental Science-Policy Platform on Biodiversity and Ecosystem Services, p. 1, 'The Platform [...] should not directly undertake new research'.

13 On this specific topic, IPCC produced a report in 2002 called 'Climate Change and Biodiversity', which can be accessed online at: www.ipcc.ch/publications_a nd_data/publications_and_data_technical_papers.shtml (last accessed: 30/06/2015).

14 Millennium Ecosystem Assessment, Ecosystem and Human Well Being, Synthesis, p. 2, available online at: www.millenniumassessment.org/documents/document.356. aspx.pdf (last accessed: 30/06/2015).

15 For more information on this topic, see: www.cbd.int/climate/intro.shtml (last accessed: 30/06/2015).

16 The Cambridge Online Dictionary gives the following definition: 'the order or method of doing something'.

17 Rules of Procedure for the Plenary of the Platform, Rule 36, available online at: http://www.ipbes.net/sites/default/files/downloads/IPBES_rules_of_procedure.pdf (last accessed: 30/06/2015).

18 Koetz, T., Farrell, K., Bridgewater, P., 'Building Better Science-policy Interfaces for International Environmental Governance: Assessing the Potential within the IPBES', *International Environmental Agreements: Politics, Law and Economics*, vol. 12, n°1, 2012, pp. 1–21, p. 3.

19 Functions, operating principles and institutional arrangements of the Inter-governmental Science-Policy Platform on Biodiversity and Ecosystem Services, p. 2, 'Be scientifically independent and ensure credibility, relevance and legitimacy through peer review of its work and transparency in its decision-making processes'.

20 Koetz, T., Farrell, K., Bridgewater, P., 'Building Better Science-policy Interfaces for International Environmental Governance: Assessing the Potential within the IPBES', op. cit., p.3, 'credibility reflects the perceived validity of information, methods and procedures provided and applied via a Science Policy Interface'.

21 *Idem*, 'relevance reflects the extent to which the work carried out within a SPI is responsive to the conditions and needs of the policy process'.

22 Ovodenko, A., Keohane, R., 'Institutional Diffusion in International Environmental Affairs', *Institutional Affairs*, vol. 88, n°3, 2011, pp. 523–541, p. 524.

23 Ibid., p. 541.

24 *Idem.*

25 See Beck, S. et al., 'Towards a Reflexive Turn in the Governance of Global Environmental Expertise: The Case of the IPCC and the IPBES', *Gaya*, vol.23, n°2, 2014, pp. 80–87.

26 Earth Negotiation Bulletin (ENB), vol. 158, n°1, 2008.

27 UNEP/IPBES/3/INF/5, Background document on the Intergovernmental Panel on Climate Change.

28 Agrawala (S.), 'Structural and Process History of the IPCC', op. cit.

29 Appendix A to the Principles Governing IPCC Work, Procedures for the pre-paration, review, acceptance, adoption approval and publication of IPCC reports, online at: www.ipcc.ch/organization/organization_procedures.shtml (last accessed: 30/06/2015).

30 IPBES/3/L.2, Procedures for the preparation of Platform deliverables.

31 Inter Academic Council, 'Climate Change Assessments Review of the Processes and Procedures of the IPCC', IAC Secretariat, Amsterdam, 2010, p. 104.

32 See Report of the 33rd Session of the IPCC, Abu Dhabi, United Arab Emirates, 10–13 May 2011, p. 4.

33 Hochkirch, A., McGowan, P., Van Der Sluijs, J., 'Biodiversity Reports Need Author Rules', *Nature*, vol. 516, 2014, p. 170.

34 Larigauderie, A., 'Pollinator Assessment: IPBES Responds on Conflicts of Interest', *Nature*, vol. 517, 2015, p. 271.

35 IPBES/3/L.6, Conflict of interest policy and implementation procedures, p. 8.

36 IPCC Conflict of interest policy, online at: www.ipcc.ch/organization/organiza tion_procedures.shtml (last accessed: 30/06/2015).

37 IPBES/3/L.6, Conflict of interest policy and implementation procedures.
38 Appendix A to the Principles Governing IPCC Work, Procedures for the pre-paration, review, acceptance, adoption approval and publication of IPCC reports, section 4.1.
39 IPCC, *Climate Change 2014: Synthesis Report. Contribution of Working Groups I, II and III to the Fifth Assessment Report of the Intergovernmental Panel on Climate Change*, Core Writing Team, Pachauri, R.K. and Meyer, L.A. eds, IPCC, Geneva, Switzerland, 2014, 151 p.
40 IPBES-1/3, Procedure for receiving and prioritizing requests put to the Platform.
41 IPBES-2/5, Work programme for the period 2014–2018.
42 See IPBES/2/INF/9, Supporting documentation on the prioritization of requests, inputs and suggestions put to the Intergovernmental Science-Policy Platform on Biodiversity and Ecosystem Services.
43 IPBES/3/L. 15, Revised stakeholder engagement strategy.
44 However, stakeholders are left to their own devices to establish the open-ended network. Because of their number and the absence of a clear and centralized centre of organization, it is hard to determine at this point what the open-ended network will look like.
45 IPBES/3/ L.8, Strategic partnerships.
46 IPBES/3/INF/13, Report on the institutional arrangements established to operationalize technical support.
47 As illustrated by the discussion on this aspect within IPCC. See IPCC-XLI/Doc. 4, Consideration of recommendations by the task force group on Future Work of IPCC, p. 5.
48 See Memorandum of Cooperation between the Secretariat of the Convention on Biological Diversity and the Secretariat of the Intergovernmental Science-Policy Platform on Biodiversity and Ecosystem Services, Pyeong Chang, 9 October 2014, available online at: www.cbd.int/doc/agreements/agmt-ipbes-2014-10-09-mou-en.pdf (last accessed: 30/06/2015).
49 See IPCC-XLI/Doc. 5, Chairman's Vision Paper on the Future of the IPCC.
50 Scott, K., 'International Environmental Governance: Managing Fragmentation through Institutional Means', *Melbourne Journal of International Law*, vol. 12, 2011, pp. 177–216, p. 192.
51 Pesche, D. et al., 'Le Consensus d'Antalya: les avancées de la Plateforme Intergouvernementale scientifique et politique sur la biodiversité et les services écosystémiques (IPBES)', *Nature Science et Société*, vol. 22, n°3, 2014, pp. 240–246.
52 Appendix A to the Principles Governing IPCC Work, Procedures for the pre-paration, review, acceptance, adoption approval and publication of IPCC reports, section 4.3.2.
53 IPBES-2/2, Multidisciplinary Expert Panel.
54 IPBES-2/3, Procedures for the preparation of the Platform's deliverables, section 3.6.2.
55 IPBES/3/2, Implementation of the Work Programme for 2014–2018, p.4, 'Con-sidering revisiting the requirement of 80% of selected experts having to come from government nominations to make it a less stringent requirement. This would potentially reduce the need for requesting governments for additional nominations.'
56 IPBES/1/INF/15, Preliminary review of the motivations for participating in Platform assessments.
57 IPBES/2/4, Conceptual Framework for the Intergovernmental Science-Policy Plat-form for Biodiversity and Ecosystem Services. For a detailed analysis, see Diaz, S. et al., 'A Rosetta Stone for Nature's Benefits to People', *PLOS Biology*, vol.13, n°1, 2015, pp. 1–8.
58 Brooks, T.M., Lamoreux, J.F., Soberon, J., 'IPBES≠IPCC', op. cit.

59 IPCC-XLI/Doc. 4, Consideration of recommendations by the task force group on Future Work of the IPCC, p. 5.
60 *Earth Negotiation Bulletin*, vol. 12, n° 67, 2015.

References

Agrawala, S., 'Structural and Process History of the IPCC', *Climate Change*, vol. 39, 1998, pp. 621–642.

Beck, S. et al., 'Towards a Reflexive Turn in the Governance of Global Environmental Expertise: The Case of the IPCC and the IPBES', *Gaya*, vol. 23, n°2, 2014, pp. 80–87.

Brooks, T.M., Lamoreux, J.F., Soberon, J., 'IPBES≠IPCC', *Trends in Ecology and Evolution*, vol. 29, n°10, 2014, pp. 543–545.

Diaz, S. et al., 'A Rosetta Stone for Nature's Benefits to People', *PLOS Biology*, vol. 13, n°1, 2015, pp. 1–8.

Hochkirch, A., McGowan, P., Van Der SluijsJ., 'Biodiversity Reports Need Author Rules', *Nature*, vol. 516, 2014, p. 170.

IPCC, *Climate Change 2014: Synthesis Report. Contribution of Working Groups I, II and III to the Fifth Assessment Report of the Intergovernmental Panel on Climate Change*, Core Writing Team, Pachauri, R.K. and Meyer, L.A. eds, IPCC, Geneva, Switzerland, 2014.

Koetz, T., Farrell, K., Bridgewater, P., Building Better Sscience-policy Interfaces for International Environmental Governance: Assessing the Potential within the IPBES, *International Environmental Agreements: Politics, Law and Economics*, vol. 12, n°1, 2012, pp. 1–21.

Larigauderie, A., 'Pollinator Assessment: IPBES Responds on Conflicts of Interest', *Nature*, vol. 517, 2015, p. 271.

Maljean-Dubois, S. 'La Plateforme Intergouvernementale Scientifique et Politique sur la Biodiversité et les Services Ecosystémiques (IPBES)', *Journal International de Bioéthique*, vol. 35, n°1, 2014, pp. 55–73.

Orsini, A., Morin, J.F., Young, O., 'Regime Complexes: A Buzz, a Boom, or a Boost for Global Governance?', *Global Governance*, vol. 19, 2013, pp. 27–39.

Ovodenko, A., Keohane, R., 'Institutional Diffusion in International Environmental Affairs', *Institutional Affairs*, vol. 88, n°3, 2011, pp. 523–541.

Pesche, D. et al., 'Le Consensus d'Antalya: les avancées de la Plateforme Intergouvernementale scientifique et politique sur la biodiversité et les services écosystémiques (IPBES)', *Nature Science et Société*, vol. 22, n°3, 2014, pp. 240–246.

Rausitalia, K., Victor, D., 'The Regime Complex for Plant Genetic Resources', *International Organization*, vol. 58, n°2, 2004, pp. 277–309.

Scott, K., 'International Environmental Governance: Managing Fragmentation through Institutional Means', *Melbourne Journal of International Law*, vol. 12, 2011, pp. 177–216.

6 The IPCC experience and lessons for IPBES

Daniel Compagnon and Wolfgang Cramer

Large-scale environmental assessments involve the participation of many scientific disciplines, even more so when repeated throughout the decades and when implemented in an institutionalized context such as the Intergovernmental Panel on Climate Change (IPCC) and the Intergovernmental Platform on Biodiversity and Ecosystem Services (IPBES). Although the purported function of these international organizations is to provide a working interface between science and decision-making circles at the international level, the "boundary work" concerns as much the coordination between scientific disciplines not used to work together and lacking a common language, as it defines interactions between the world of politics and the scientific community. The development of common standards for scientific evidence and plausibility is crucial, both within and between the natural and social sciences. The challenge is to produce statements of greater clarity than otherwise found in scientific publications for the use of policy makers, without jeopardizing the established scientific standards.

The institutional similarities between these two international organizations are deliberate and derived from the same willingness of UN member states to retain some control over potentially far-reaching scientific assessments (see chapter 2). Institutional differences are largely a consequence of more than two decades of experience between the establishment of IPCC and that of IPBES, and the creation of a more favorable political climate during this time: Today, many member states more clearly recognize the potential for valuable policy support provided by science-policy interfaces as compared to the 1980s. Likewise, various lobbies acknowledge the importance of an institutionalized science-policy interface in the global policy debate on climate change. The development of IPBES will hopefully benefit from the IPCC experience in two areas, i.e. interaction with policy makers and complex scientific coordination. Therefore, it is important to understand the nature of IPCC and its limits and shortcomings, and also to understand the differences between the two institutions.

The first section of this chapter reviews the social sciences literature devoted to IPCC and its reports without endorsing all the opinions expressed, while reflecting on the kinds of problems and challenges facing IPBES. The second section, based on the personal experience of the second co-author, both in IPCC and in the nascent IPBES, illustrates these problems and emphasizes the specificities of IPBES in its early stages.

Ambiguities of the IPCC "model"

Originally seen as an obscure specialized institution, the International Panel on Climate Change (IPCC) has become an iconic example of a strong influence of science in the construction of international environmental regimes, even though this picture is not totally accurate. Its public opinion status was greatly enhanced by being awarded the 2007 Nobel Peace Prize (shared with Al Gore). There is considerable literature on IPCC, only part of which is in the fields of political science/international relations (for a review, see Hulme & Mahony, 2010).

IPCC was established in November 1988, on the request of the G8 countries, under the authority of the World Meteorological Organization (WMO) and the United Nations Environmental Programme (UNEP) – a move that was condoned by UN member states in the General Assembly Resolution 43/53 of 6 December 1988. The Toronto conference organized by UNEP in June 1988, when discussions about IPCC were under way, underlined both the importance of the issue and the lack of scientific consensus (particularly within the US scientific establishment) on both the reality and causes of climate change. It was the outcome of a decade of attempts to tackle the scientific challenges of climate change, in particular the World Climate Programme (WCP), a research initiative under WMO, UNEP and the International Council of Scientific Unions (ICSU) that was set up in 1979.

The intergovernmental nature of the new organization underlined that many governments wanted to exert some effective control over the production/dissemination of credible, legitimate policy-relevant knowledge on climate change. The conservative US and British governments in particular did not trust UNEP, which they perceived as partial to NGOs and developing countries, and whose Executive Director Mostafa Tolba had been a strong advocate of a climate convention since 1985 (Agrawala, 1998). As the different US government agencies were divided on the issue (especially the Department of Energy vs. the Environmental Protection Agency), and huge business interests (fossil fuel related industries) were at stake, some US decision makers saw it expedient to claim that acquiring more knowledge was a prerequisite for action. To them, acting on the basis of the precautionary principle (such as was the case in the Ozone Layer protection regime) was not an option. Establishing an international assessment mechanism under UN control could serve to postpone policy action until the scientific basis was deemed robust enough. Others who supported the creation of IPCC hoped, on the contrary, that an independent, scientifically grounded organization would counterbalance the influence of business lobbies in policymaking and stimulate international policy cooperation.

In this context, the definition of the "science" of climate change was also a political question from the outset, although the organization's motto made it clear that its work was "policy-relevant and yet policy-neutral, never policy-prescriptive". It is quite ironic that, in the late 1990s and early 2000s, IPCC was accused by critics from the neo-conservative and climate denialist camp of having "politicized" the science of climate change. It is the intergovernmental nature of IPCC that makes its assessments acceptable as a basis for policymaking by government negotiators in United Nation Framework Convention for Climate Change (UNFCCC). "However, having an intergovernmental status has imposed significant costs also: IPCC assessment summaries are widely regarded as being politically negotiated, which has, at times, undermined their credibility" (Agrawala, 1998: 611). Instead of just "speaking truth to power", the traditional function devoted to science (Price, 1965), climate scientists are involved in a political role through a process that produces – for the sake of policy relevance – what some people call "imperfect science" (Malnes, 2006).

IPCC is indeed an intergovernmental cooperation organization with nearly universal membership of 195 countries, and the Panel Plenary is the core decision-making body attended by government representatives. Since the Panel Plenary elects the IPCC Chairperson and the members of the Bureau (including the Co-Chairs of the three working groups and the task force on greenhouse gas inventories) from amongst candidates proposed by member states, the whole IPCC structure is kept under close supervision of the states.[1] The fact that all these officials have a robust scientific track record does not diminish the influence of the states. In addition, the Plenary also adopts the structure and mandate of IPCC Working Groups and Task Forces, the IPCC work plan, the scope and outline of IPCC reports and it rather formally approves, at the end of the assessment process, the Working Group reports and the Synthesis reports, and the respective Summaries for policymakers on a line-by-line basis, until a consensus is reached. Therefore, the intervention of the member states is by no means limited to the final phase (see Figure 6.1). IPCC works on a dual scientific and political consensus basis.

The USA, which had played an important role in the WCP, had accumulated the largest body of knowledge on climate change in the 1970s and 1980s through its national research programs, and was also then the leading emitter of greenhouse gases on a per country and per capita basis. The US government kept a close eye on the IPCC proceedings. In 2002, for example, the US government pressed for the replacement of British scientist Robert Watson, IPCC Chairperson since 1996 and a former Associate Director for Environment in the Office of the US President, who was perceived as a *bête noire* of the US oil industry by the Bush administration and too outspoken on the anthropogenic causes of climate change (MacKenzie, 2002). The US government also objected to what it considered to be politically charged language in the Third Assessment Report (2001) (Haas, 2005: 394). Governments

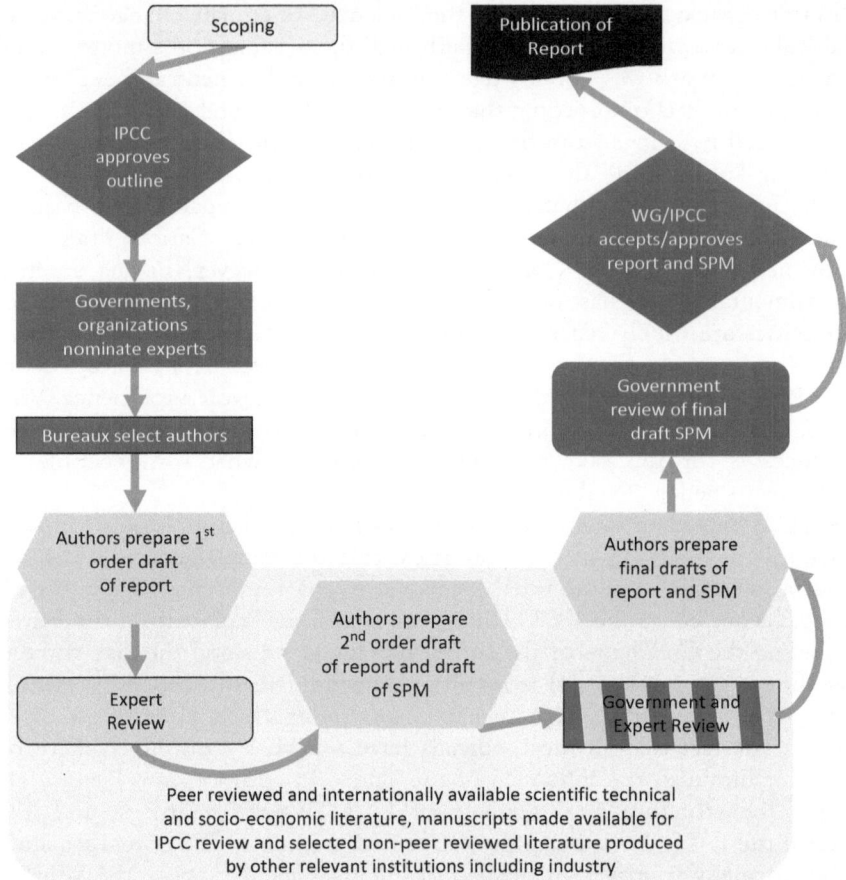

Figure 6.1 IPCC process from scoping to publication of reports (scoping is the pre-
         paratory meeting that proposes a scope and outline for the future report;
         see www.ipcc.ch/index.htm)
Note: SPM – summary for policy makers

from Russia and Saudi Arabia criticized the IPCC reports when seen as
undermining their interests, but were opposed by other (notably European)
governments who insisted on the clear language and scientific credibility of
these reports. These dynamics contributed to the politicization of the process.

Some analysts see the control of states over the IPCC organization and the
contents of the assessment reports as diminishing over the years (Siebenhüner,
2003). Subsequent successive reports have gained in credibility due to the
implication of a growing community of scientists in the assessments and the
writing of the three working group reports. IPCC produces no new research
but collects material published in scientific and publicly accessible, mostly
peer reviewed journals and books, and to a lesser extent from other sources,
evaluated according to their relevance and scientific credibility. It also

improved its internal procedures to control errors following the controversy that developed over some errors contained in the Fourth Assessment Report (2007). The Inter-Academy Council was tasked to audit the IPCC procedures and make some recommendations in a politically loaded context with the release of climate scientists' e-mails (the "climategate" affair) just before Copenhagen in 2009. Following its report in August 2010 (IAC, 2010), a number of significant improvements to the IPCC procedures were introduced, although the main structures were maintained as they were. The reform stopped short of addressing "IPCC's relationship to public policy and to its various global "publics" [and] its normative commitments in terms of accountability, political representation, and legitimacy" (Beck et al., 2014: 82).

Indeed, the legitimacy of IPCC is still contested by critics who emphasize the over-representation of scientists from the affluent West. Being a G8 recommendation, the establishment of IPCC was originally seen with suspicion by developing countries, which were concerned that an organization dominated by Western scientists would foster the political and economic interests of industrialized countries. Despite the Chair's and Secretariat's efforts to increase involvement of experts from the Global South and the election of an Indian Chairperson (Rajendra Kumar Pachauri) in 2002, and again in 2008, the proportion of OECD/non-OECD authors and reviewers remained stable through the years (Haas, 2005: 395; Hulme & Mahony, 2010). Another apparent weakness of the IPCC process is the perceived bias against social sciences, regarded as lacking the rigor of the hard sciences, with the exception of economics where quantitative assessments are used in a way that can also be understood by climate scientists (Bjurström & Polk, 2011). Some observers see the same trend developing in the nascent IPBES (Morin et al., 2015). Many of the climate scientists who have contributed to the IPCC reports are aware of these shortcomings, but the production of interdisciplinary knowledge is hampered by acute epistemological and methodological problems.

Social sciences still debate over the characterization of IPCC – a scientific lobby, a political organization or both, an epistemic community or a "boundary" organization. The notion of "epistemic community" as defined by Peter Haas emphasizes the role of consensual, policy-relevant knowledge in regime making. However, the definition as "a transnational network of professionals with recognized expertise and competence in a particular domain and an authoritative claim to policy-relevant knowledge within that domain or policy area" (Haas, 1992) does not fit the IPCC structure. In Haas' case study – the Mediterranean Action Plan – the experts were embedded in the decision-making process and coordinated transnationally by UNEP, whereas IPCC is a formal international organization and professes "policy-neutral" and certainly "not policy prescriptive" activities.

Although the First Assessment Report (1990) is said to have played a major role in the signature of the UNFCCC, and the Second Assessment Report (1995) provided major inputs in the negotiation of the Kyoto Protocol, the publication of the reports did not closely coincide with the negotiation process, leading some observers to question their impact on policymaking (Haas, 2005: 396; Victor, 2015). IPCC's major contribution has been to establish beyond doubt the reality of climate change, its predominant anthropogenic causes and the seriousness of its expected impacts, producing a "shared scientific understanding of the climate issue" (Lidskog & Sundqvist, 2015: 12). However, it failed to shape policy in any significant way, with the exception of EU policies or those in countries like Norway, in the absence of "usable knowledge" at a scale (national and sub-national) where policy makers need to make decisions.

The goal to limit global warming at 2°C above pre-industrial levels is often presented as an outcome of its reports, however, it is not actually a IPCC recommendation at all but rather a political target. It appeared for the first time in a decision of the European Council in 1996, resurfaced in a report of the German scientific advisory institute WBGU[2] in 2007 as the limit of acceptable warming, it was then adopted by EU as a comprehensible policy objective for the public, and was pushed in negotiations until being endorsed by the Copenhagen Agreement and the Conference of the Parties (COP) 16 held in Cancun in November 2010 (Aykut & Dahan, 2011). Its centrality as a policy objective and legitimate discourse on climate change is not backed by overwhelmingly scientific evidence. Science has always, throughout all IPCC reports, recognized that significant damage has already occurred (at a global warming level of roughly 1°C), and that damage will grow incrementally.

This example underlines how difficult it might be to endorse the classification of IPCC as a "boundary organization" between the world of science and the world of politics, as understood in Science and Technologies Studies (STS). Far from being pure and neutral, science is embedded in social institutions and discourses. Scientific "truths" are relative to a social-historical context and knowledge is always contested. As scientists respond to social demands, particularly from policymakers who largely set the research agenda, policy-relevant science is always a co-production (Jasanoff, 1996), or what Bruno Latour calls "hybridization" that reflects the blurred boundaries between science and policy. "Coproduction happens when policy influences the production and stabilization of knowledge, while knowledge simultaneously supports and justifies policy" (Lidskog & Sundqvist, 2015: 6). Depending on the research findings offered, policy makers will tend to be selective about what to use and not to use. In return, IPCC has actively contributed to the construction of climate change science as policy-oriented knowledge, with certain epistemic and methodological limits.

However, if the boundary organization creates a network of relations between scientists and decisions makers, the Subsidiary Body for Scientific and Technological Advice (SBSTA) of UNFCCC qualifies better than IPCC in this respect, as it is officially tasked with translating science findings into policy advice and allowing government officials to mix with scientists on a regime implementation agenda, so it engages in "hybrid management" (Miller, 2001). Discussions on policy issues[3] take place in Convention committees such as SBSTA, as noted by the first IPCC Chairman who attended many of its meetings (Bolin, 2007) at a time when IPCC was kept separate from UNFCCC. The fact that SBSTA decided to "act as the interface between the IPCC and the climate regime, with the IPCC submitting its reports to SBSTA" (Miller, 2001: 260) and to create a joint liaison group between the two institutions, underlines that IPCC in itself is not the real science-policy interface, as often suggested. Although it is a typical UN forum following the consensus rule of procedure, SBSTA has moved at a slow pace and its proceedings were heavily politicized, while allowing greater participation from smaller countries and attention to their worldviews (Ibid.: 268).

Besides, does IPCC really sit "on the boundary"? The analysis of its structure and rules of procedure briefly sketched above rather suggests that it is fully inter-governmental, with the original intent "to keep science on a tight leash" (Haas, 2005: 396). Yet, within IPCC, we should differentiate the large community of scientists who are contributors and lead authors in the various chapters of the working group reports, and the panel of government representatives. In retrospect, it seems that through this duality – IPCC and SBSTA – the principals (governments) have carefully circumscribed the role of science in the climate change regime in order to retain substantial political decision-making autonomy. They could not, however, prevent the IPCC reports from gaining a wider audience in the media and civil society, which in turn has exerted a certain pressure on decision makers. Thus, it is through the public space (in the sense of Habermas and Dryzek) that IPCC reports have influenced policy making in a broader sense.

From the history of IPCC, there are significant lessons to be drawn for IPBES. Largely inspired by IPCC (Larigauderie & Mooney, 2010), this Platform on Biodiversity and Ecosystem Services is also an intergovernmental organization where the Plenary is the main decision-making body. It is positioned in relation to the Convention on Biological Diversity (CBD) very much in the way IPCC is positioned in relation to UNFCCC, outside the negotiation process, and IPBES relies on member states for implementation (see chapter 3). CBD also has a Subsidiary Body on Scientific, Technical and Technological Advice (SBSTTA) to provide policy-relevant advice to the parties. In addition, the status of IPBES assessments in relation to the Global Biodiversity Outlook (GBO) regularly published by the CBD Secretariat is yet to be decided. IPBES is also faced with the same challenges of involving more scientists from the Global South, or of balancing different knowledge systems. Once again, at the beginning of IPBES, some observers pointed at the quasi-absence of social scientists in the governance of the organization, when social sciences could help develop a more self-critical and reflexive approach to knowledge (Lemos & Morehouse, 2005). However, since 2015 and the more explicit integration of non-scientific knowledge systems, including local knowledge from communities playing a key role in conservation of natural resources and habitats, four social scientists are involved in the Multidisciplinary Expert Panel (MEP).[4]

The next section will explore the similarities between IPCC and IPBES from the personal experience of Wolfgang Cramer, a scientist who was involved in both.

IPCC and IPBES from within

Participating in science-policy interface work is a challenge for most scientists and can be a life-changing experience for some. Arguably, no field of science other than climate (and now biodiversity science) offers the same opportunity for intensive exchange on key issues between decision makers in international politics and the scientific community. A guiding principle is mutual respect, as prominently captured by the "policy-relevant but not policy-prescriptive" principle, which initially poses problems for some scientists who believe that it is their role to convince policy makers on the need to act. So the first lesson to learn is not only to speak clear and jargon-free language, but also to listen to and understand the professional context within which the policy maker plays his/her role.

What is seldom recognized by outsiders of the process is that researchers participate in science-policy interface work on a voluntary basis – both for IPCC and IPBES. There is no remuneration and there is the expectation that a large amount of time needs to be invested. Indeed, most scientists consider it an honor to be invited to contribute, and they invest significant amounts of their spare time in this work. Most roles are assigned by invitation, involving a selection procedure that gives preference to scientific credentials, but then also strives to involve "political" elements such as the goal to increase the participation of colleagues from low-income countries, to have an adequate gender balance and also high participation from different scientific disciplines. While this process makes a great point of departure for the building of an international expert community, it also creates an obstacle to wide participation because many scientists do not work in positions where they can, at their own discretion, allocate part of their time (during or outside of working hours) to IPCC or IPBES.

I will look, from my personal perspective, at a few of the different roles an individual scientist participating in IPCC must fulfill – many of these roles will have to be performed in IPBES too. Clearly the most accessible role is that of a reviewer. For the Fifth Assessment Report, invitations to review chapter drafts were disseminated very broadly, and any interested experts were given access to draft documents in order to provide comments on them. The role of these reviewers is hugely important for the quality of the assessment, but it provides little to no benefits to them in terms of enhanced reputation or career advantages. It is therefore often hard to motivate scientists to carefully review report drafts. Yet their contributions are increasingly important given that, due to the growing amount of published scientific knowledge, the full amount of information can really no longer be assessed by an individual report author. Even if a reviewer reads only some sections, for whatever reason related to personal interest, the contribution will be useful and appreciated by the report authors. I first reviewed draft IPCC chapters for the Second Assessment Report in 1991, spending many days on them. While reading, I learned a lot about aspects of scientific studies that I was unaware of, but I also found gaps and mistakes. My sometimes detailed responses led to an invitation to one of the author meetings, and I so became more deeply involved in the process. In particular, there was an intensive and unresolved debate within one of the author teams where my comments helped resolve the issue.

The chapter writing process has changed substantially over the past 25 years, because the amount of scientific literature keeps increasing, and the goal is still to provide a comprehensive overview of the available knowledge. Without active reviewers, this can no longer be achieved, and it is important to increase the number of experts participating in the process.

Other roles for IPCC or IPBES are usually assigned after some nomination and selection procedure. National "focal points" in most countries assist their governments in this process, usually with the aim of encouraging participation from their country in the Platform. Being a "lead author" is a quite important role. These experts are asked to summarize the knowledge for some part of the science covered by a chapter, to provide text for inclusion, and to consider the comments that have been made by the reviewers. As lead author in various chapters of the Third and Fourth Assessments of IPCC, I had the opportunity to work closer than I otherwise would with some leading scientists in my discipline, and several highly useful professional collaborations have arisen from these exchanges. Being a lead author also inevitably broadens your view of your own field of science, as you must also assess literature from topics outside of your "scientific comfort zone". Authorship at this level, as a member of a small international team, is a great opportunity to sharpen one's abilities to identify "key findings" and present them in a way that communicates them clearly to non-specialists. During the construction of the overall assessment, one can then watch and be involved in the process of synthesizing key findings into an overall message that may later be found in the summaries of the chapter, the working group volume or even the overall assessment report.

A basic but ongoing problem for authors is the limitation in the space they are given. Authors usually find their topic extremely important and they struggle to provide text that contains precise information while still being short enough to be accessible to readers who typically look for information on many topics at the same time. Another problem is that "assessment" implies not only reporting facts that are documented in the literature, but also providing statements on their reliability. In ecosystems, for example, species often come and go due to natural fluctuations, to direct human interference or to climate change. In an assessment of these changes, the author is expected to provide clear guidance on the relative role for these different "drivers". When there is no definitive answer (a situation that frequently

arises), then the author must communicate just how uncertain a given conclusion is, considering all available evidence. IPCC produces fairly detailed
guidance documents to help authors achieve this task. In recent reports,
authors have used what is referred to as "calibrated uncertainty language", a
useful tool but one that requires training and therefore additional time
investment for authors.

Being a "coordinating lead author" is a particularly significant IPCC task –
I had this role in the recent (fifth) assessment, co-leading a chapter on the
"detection and attribution of observed impacts" for Working Group II
(Cramer et al., 2014). The coordinating task obviously implies the maintenance
of a good communication process with the lead author's team, which by
necessity is a global team, involving colleagues from similar disciplines but
different geographical and cultural backgrounds, in our case from Brazil,
Germany, the Philippines, South Africa, Sweden and the USA. Given the
great distances, such teams can physically meet only rarely, i.e. in our case
only five times during the six years of work, at quasi-annual meetings, so it is
imperative that good electronic communications are put to use.

In our specific case, we identified a stronger-than-usual necessity to interact
not only with each other but also with the authors of a large number of other
chapters under preparation for the same report. This was because, in different
regions and different scientific domains, the notion of what constitutes a
scientifically plausible detection of some climate change impact from observations may differ. To ensure consistency, it was necessary to first develop
clearer terminology and then train colleagues from different domains and
regions in its use. Our objective was that all chapters treating regional issues
or specific domains (like ecosystems, agriculture, settlement and others)
would use the same detection and attribution methods, so that each chapter
would provide a synthesis of the findings outlined in all the other chapters.
To achieve this goal, we created a "social network", connecting members of
our own team electronically to contact persons involved in the 20 sectoral/
regional chapters of the same report. With these individuals, we not only
exchanged emails, but we also organized web conferences where we discussed our approach and its consequences for the assessment. During one
author's meeting, in Buenos Aires (Argentina), we even booked a whole
restaurant in order to strengthen our social links to the other authors, and
this turned out to be successful.

What emerged from this experience of being a coordinating lead author is,
firstly, that the task is much beyond the editorial effort of aggregating pieces
of text into a coherent document. The early definition of the shared objectives,
and continuous monitoring of the process are crucial, internally for the
chapter but also for the connection between the chapter and all others in the
same working group report. The task therefore becomes a social process
demanding a substantial effort from the coordinator, as well as from everyone else involved. Increasingly, governments recognize that coordinating lead
authors need support for carrying out this task due to the volume of

available information. In our case, with support from the German govern‐ment, the work was supported by an expert in her own right on climate change issues, who worked full‐time for the chapter team – without her, the result would have been a mere shadow of what the chapter is now.

IPCC reports strive to achieve accessible syntheses by adding executive summaries to the chapters, and a technical summary and a summary for policy makers to the working group reports. Preparation of the latter repre‐sents a considerable challenge for the coordinating lead authors as they must engage delegations of all governments in a one week‐long event called the "approval plenary". During this session, the full draft of the summary is projected on large screens, and every sentence of it is open to questions from government delegates. For any statement relating to a particular chapter, the coordinating lead author must be ready to explain the underlying scientific evidence. While the situation itself may feel like an academic examination (something most of us had thought we would never again have to face in our professional life), this is also a satisfying opportunity to explain the findings, to seek understanding for them, and, if necessary, to modify the draft to further improve its clarity. It has sometimes been claimed that this is a "negotiation" situation, and it is true that one cannot help but feel that some governments raise issues with the sole purpose of weakening some statement. But the principles are clear and undisputable – the report is only under the responsibility of the scientists, and changes are made with the principal objective of reaching an unambiguous and scientifically sound text. There is simply no mechanism that could introduce, upon government request, any statement that could be scientifically incorrect. Text is sometimes deleted by the scientists chairing the discussion because a significant number of govern‐ment delegations consider that it could lead to misunderstandings on critical issues.

As coordinating lead author for the chapter on the detection and attribution of impacts of recent climate change events that have already occurred, I had to deal with an unexpected situation during the approval plenary for the IPCC's Fifth Assessment Report, in Yokohama in March 2014, while presenting a global map of observed impacts of climate change.

Since this map could only document cases where some impacts had not only been observed but also attributed to climate change on the basis of a published scientific study, it was evident to all parties in the room that many impacts must have occurred that are not actually shown on our map. This of course was no omission, it was a clear consequence of the scientific method being applied to any part of the IPCC assessment. For a number of country delegations, this nevertheless appeared to be misleading since the map apparently showed "no impacts" where in fact impacts were likely to have occurred. For some time I tried to meet these concerns but failed to con‐vince the delegates. The IPCC Chairman then decided to call an informal meeting, in the corridors of the conference center, with these delegations, myself and any other interested parties. As a result of this constructive

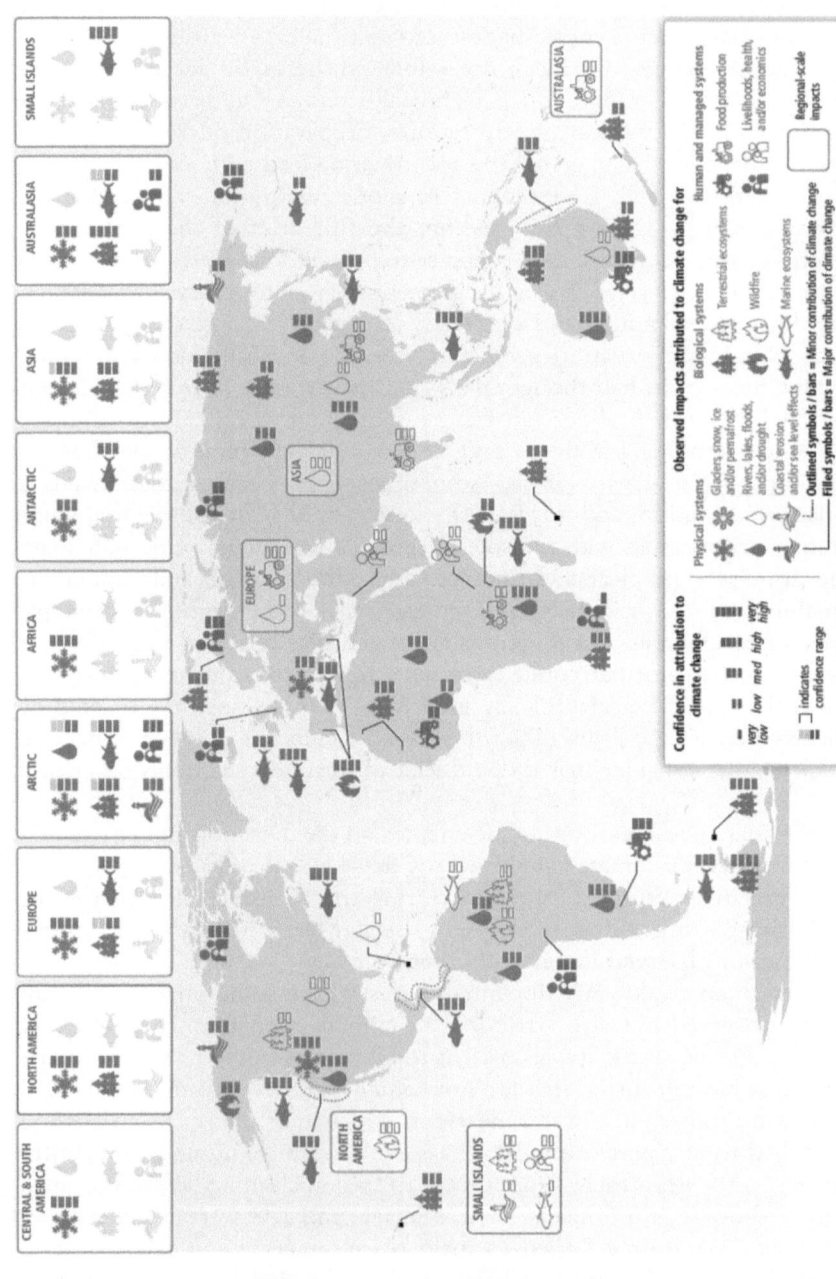

Figure 6.2 Global patterns of observed climate change impacts reports since Assessment Report 4

Source: Cramer et al. (2014)

discussion in the corridor, we further strengthened the statements made in the explanatory text of the map, such as to make sure that no erroneous conclusions on the absence of observed impacts could be drawn from it. For me, this exchange was personally rewarding. It became clear to me that government delegates were very serious and wanted clear and unambiguous messages, and that the concern was that some important impacts might end up not being recognized.

A further step in the drafting of clear IPCC statements from scientific evidence is the development of a separate Synthesis Report, aimed at pooling the most important conclusions from the three working group reports into a shorter document. Participating, as I did, in the work of the core writing team for this report is also highly rewarding. This is because, in an even broader setting, one has to face the challenge of debating with colleagues the relationship between different findings and their importance for policy making (specific aspects of this process, still concerning the map from our chapter, are outlined in a commentary by Hansen & Cramer, 2015).

In closing, I would like to mention a last organizing principle for IPCC and IPBES assessment work. Throughout the various layers of summary and synthesis writing, it is extremely important that any top-level statement can be traced back to the original scientific publication and even its underlying data. In principle, this is ensured by appropriate references, and all quality control steps need to verify that no errors or omissions occur during the editorial process. In the case of a heavy political debate that arose about the correct interpretation of some summary findings of the Fourth Assessment Report, concerning the numbers of people potentially affected by agricultural yield losses in Africa, we actually documented the "line of sight" from the summary statement back to the underlying scientific reference in detail (Müller et al., 2011). This was useful for two reasons. First, it demonstrated that no actual errors had been made during the assessment and subsequent synthesis. And second, it also highlighted that the underlying scientific information urgently requires a greater research effort in the future.

Ultimately, despite the growing availability of scientific information, it is undisputable that our knowledge is limited. While in the case of changing climatic conditions and declining biodiversity, the time is well overdue to take much stronger action for preservation of the environment; it also demonstrates that a stronger research effort is needed for better adaptation to the inevitable future changes. Being part of this research *and* communication effort has been a very enriching experience in my professional and personal life.

Conclusion

Both the 27 years of IPCC existence and the more recent IPBES experience emphasize the importance of inter-scientific dialogue and the necessity to broaden the scope beyond the core natural sciences. Although this objective is largely shared by a great majority of scientists and some significant cooperation has been achieved so far, there is room for improvement still. In particular, social sciences (or local and indigenous knowledge for that matter) are not there just to inform policy once the diagnosis has been produced by hard sciences, they must also be part of the definition of the problem from the outset.

IPBES will benefit most from the IPCC experiment if it encourages the circulation of scientists between the two processes. Robert Watson's involvement in the IPBES structure is a case in point, but the participation of authors and lead authors such as the second author of this chapter are equally important. They will not only facilitate linkages between the two series of assessments for the many overlapping issues between climate change and biodiversity loss. While IPBES certainly should avoid making the same errors as IPCC, some memory from previous experiences will still be useful.

Finally, the history of building these assessment processes through a trial and error approach shows that the debate on neutrality of "pure" science vs. politicization of science has not been very productive. Both totally independent assessments and embedded scientific bodies can succeed or fail in making "science speak the truth to power". As social constructs, these assessment bodies will always be challenged and criticized, but such debates should not undermine the bodies' legitimacy, hence reducing their actual capacity to influence policy. This is particularly important for the IPBES future, when strong policy commitments are expected from megadiverse countries, some of which were initially skeptical about the establishment of the Platform altogether.

Notes

1 The original division in three working groups was proposed by UNEP Executive Director Mostafa Tolba, i.e. a tradeoff owing more to politics than to science (Bolin, 2007: 49–50).
2 Wissenschaftliche Beirat der Bundesregierung Globale Umweltveränderungen.
3 SBSTA focuses on issues other than climate risk assessment, such as technology transfers, standards and methodologies for measuring emissions and abatement efforts.
4 In 2015, four of the 25 MEP members were social scientists (Marie Roué, [anthropology], Unai Pascual [economics], Marie Stenseke [human geography] and György Pataki [management and communication]).

References

Agrawala, S. 1998. "Context and Early Origins of the Intergovernmental Panel on Climate Change." *Climatic Change*, 39, 605–620.
Aykut, S.C. & Dahan, A. 2011. "Le régime climatique avant et après Copenhague: sciences, politiques et l'objectif des deux degrés." *Natures, Sciences et Sociétés*, 19(2), 144–157.

Beck, S., Borie, M., Chilvers, J.Esguerra, A., Heubach, K., Hulme, M., Lidskog, R., Lövbrand, E., Marquard, E., Miller, C., Nadim, T.Neßhöver, C., Settele, J., Turnhout, E., Vasileiadou, E. & Görg, C. 2014. "Towards a Reflexive Turn in the Governance of Global Environmental Expertise. The Cases of the IPCC and the IPBES." *GAIA: Ecological Perspectives for Science and Society*, 23(2), 80–87.

Bjurström, A. & Polk, M. 2011, "Physical and Economic Bias in Climate Change Research: A Scientometric Study of IPCC Third Assessment Report." *Climatic Change*, 108(1–2), 1–22.

Bolin, B. 2007. *A History of the Science and Politics of Climate Change: The Role of the Intergovernmental Panel on Climate Change*. Cambridge (UK): Cambridge University Press.

Cramer, W., Yohe, G., Auffhammer, G., Huggel, C., Molau, U., Silva Dias, M.A.F., Solow, A., Stone, D. & Tibig, L. 2014. "Detection and Attribution of Observed Impacts." In Field, C.B., Barros, V.R., Dokken, D.J., Mach, K.J., Mastrandrea, M. D., Bilir, T.E., Chatterjee, M., Ebi, K.L., Estrada, Y.O., Genova, R.C., Girma, B., Kissel, E.S., Levy, A.N., MacCracken, S., Mastrandrea, P.R. & White, L.L. (eds.) *Climate Change 2014: Impacts, Adaptation, and Vulnerability. Part A: Global and Sectoral Aspects. Contribution of Working Group II to the Fifth Assessment Report of the Intergovernmental Panel on Climate Change*. Cambridge (UK) & New York (NY): Cambridge University Press, 979–1037.

Haas, P.M. 1992. "Introduction: Epistemic Communities and International Policy Coordination." *International Organization*, "Knowledge, Power, and International Policy", 46(1), 1–35.

Haas, P.M. 2005. "Science and International Environmental Governance." In Dauvergne, P. (ed.), *Handbook of Global Environmental Politics*. Cheltenham UK: Edward Elgar, 383–401.

Hansen, G. & Cramer, W. 2015. "Global Distribution of Observed Climate Change Impacts." *Nature Climate Change*, 5, 182–185, doi: 10.1038/nclimate2529.

Hulme, M. & Mahony, M. 2010. "Climate Change: What Do we Know about the IPCC?" *Progress in Physical Geography*, 34(5), 705–718.

Jasanoff, S. 1996. "Science and Norms in Global Environmental Regimes." In Hampson, F.O. & Reppy, J. (eds), *Earthly Goods: Environmental Change and Social Justice*, Ithaca, NY: Cornell University Press, 173–197.

IAC (InterAcademy Council). 2010. *Climate Change Assessments: Review of the Processes and Procedures of the IPCC*. Amsterdam: IAC.

Larigauderie, A. & Mooney, H.A. 2010. "The Intergovernmental Science-Policy Platform on Biodiversity and Ecosystem Services: Moving a Step Closer to an IPCC-Like Mechanism for Biodiversity." *Current Opinion in Environmental Sustainability*, 2, 9–14.

Lemos, M.C. & Morehouse, B.J. 2005. "The Co-Production of Science and Policy in Integrated Climate Assessments." *Global Environmental Change*, 15(1), 57–68.

Lidskog, R. & Sundqvist, G. 2015. "When Does Science Matter? International Relations Meets Science and Technology Studies." *Global Environmental Politics*, 15(1), 1–20.

Mackenzie, D. (2002). "Too Hot for Head of Climate Panel". *NewScientist.com*. 20 April. https://www.newscientist.com/article/mg17423392-400-too-hot-for-head-of-climate-panel/. Consulted on 23 July 2015.

Malnes, R. 2006. "Imperfect Science." *Global Environmental Politics*, 6(3), 58–71.

Miller, C.A. 2001. "Challenges in the Application of Science to Global Affairs: Contingency, Trust, and Moral Order." in Miller, C.A. & Edwards, P.N. (eds.),

Changing the Atmosphere: Expert Knowledge and Environmental Governance, Cambridge (MA): MIT Press, 247–286.

Morin, J.-F., Louafi, S., Orsini, A., Oubenal, M. 2015. *Boundary Organizations in Regime Complexes. A Social Network Assessment of IPBES*, New Orleans: International Studies Association.

Müller, C., Cramer, W., Hare, W.L. & Lotze-Campen, H. 2011. "Climate Change Risks for African Agriculture." *PNAS*, 108(11), 4313–4315, doi: 10.1073/pnas.1015078108.

Price, D. 1965. *The Scientific Estate*. Cambridge, MA: Harvard University Press.

Siebenhüner, B. 2003. "The Changing Role of Nation States in International Environmental Assessments – The Case of the IPCC." *Global Environmental Change*, 13, 113–123.

Stone, D., Auffhammer, M., Carey, M., Hansen, G., Huggel, C., Cramer, W., Lobell, D., Molau, U., Solow, A., Tibig, L., Yohe, G. 2013. "The Challenge to Detect and Attribute Effects of Climate Change on Human and Natural Systems." *Climate Change*, 121(2), 381–395, doi: 10.1007/s10584-10013-0873-0876.

Victor, D.G. 2015. "Embed the Social Sciences in Climate Policy." *Nature*, Comment, 520, 27–29.

7 Making the IPBES conceptual framework

A Rosetta Stone?

Maud Borie and Denis Pesche

At the international level, institutions of global expert advice responsible for the provision of policy-relevant knowledge are in increasing demand. In particular, since the creation of the Assessment panel on Ozone in the early 1980s, global environmental assessments (GEAs) have become increasingly relied upon with the most prominent example being the Intergovernmental Panel on Climate Change (IPCC, 1988) and, more recently, IPBES. Mitchell and colleagues define assessments "as formal efforts to assemble selected knowledge with a view toward making it publicly available in a form intended to be useful for decision making" (Mitchell et al., 2006: 3). Underlying these organizations is the assumption that scientific knowledge and experts have a key role to play to address pressing environmental issues such as climate change or biodiversity loss. In doing so, GEAs contribute to the definition of the "problem" at stake while delineating whose knowledge and expertise should be included in conducting assessments. Implicitly or explicitly, they operate with particular conceptual frameworks which play an important role in delineating what counts as useful or policy-relevant knowledge. According to its own terms, IPBES aims to "strengthen the science-policy interface for biodiversity and ecosystem services for the conservation and sustainable use of biodiversity, long-term human well-being and sustainable development".[1] In order to address this ambitious objective, IPBES has adopted a conceptual framework to be used in all functions of its mandate, currently organized around four functions including providing: (1) assessments, (2) policy-support tools, (3) knowledge generation and (4) capacity-building.

Science-policy interface (SPI) research has been flourishing. SPIs are generally conceived as "social processes which encompass relations between scientists and other actors in the policy process, and which allow for exchanges, co-evolution, and joint construction of knowledge with the aim of enriching decision-making" (van den Hove, 2007: 824). Organizations such as IPBES, which contribute to the organization and management of science-policy relations are often referred to as "boundary organizations" (Guston, 2001). This concept underlines the hybrid nature of these institutions which bring together heterogeneous actors from both the "science" and "policy" worlds (e.g. scientists, experts, civil servants). It also emphasizes that boundaries between science and policy are not given in advance and builds on Gieryn's concept of "boundary work" (1983) to scrutinize the ways in which such boundaries are institutionalized. It is therefore an invitation to study the specific ways in which knowledge is constructed as "policy-relevant". Far from seeing science and policy as being linked with linear processes, literature on boundary organizations stresses that science can never unconditionally speak "truth to power" (e.g. Koetz et al., 2012).

So far, much literature on GEAs has focused on analyzing how these organizations can achieve "credibility, relevance and legitimacy" (e.g. Clark et al., 2006; Farrell et al., 2001), and these criteria have become guiding objectives for IPBES. But what happens when a fundamental tension emerges about the content and the concepts themselves? This chapter presents the process leading to the development of the IPBES conceptual framework (CF) as well as the content of this framework. It focuses particularly on the main controversies that have characterized its development, between experts, governmental delegations, and other IPBES stakeholders. The goal of the IPBES CF is to provide a common heuristic tool to facilitate the participation of the diverse actors participating in IPBES and coordinate their activities. Yet the process has been animated by numerous debates and politico-epistemic struggles – the development of this framework was controversial and diverse discourses regarding how to articulate relations between nature and society were highlighted.

In analyzing the role of discourses in policy processes, Vivien Schmidt distinguishes two kinds of discourses: coordinative discourse among "policy actors engaged in creating, arguing, bargaining, and reaching agreement on public policies in the policy sphere" and the "communicative" discourse "between political actors and the public engaged in presenting, contesting, deliberating, and legitimating such policies. Coordinative discourse led to discursive communities and communicative discourse is framed for the broader public" (Schmidt, 2008). In our case, IPBES CF can be conceived as material support for "coordinative discourse", with the aim of creating a shared vision and acting as a standardization device. Following its adoption, IPBES CF was referred to as a "Rosetta Stone", a metaphor which suggests that this framework can also serve as an instrument to perform a form of "communicative discourse". In other words, beyond its coordinative role, IPBES CF is also used to construct the public identity of IPBES and ensure its credibility in front of multiple audiences. From a symbolic standpoint (e.g. Hilgartner, 2004), IPBES CF can therefore also be seen as an important element through which IPBES is attempting to achieve global credibility.

The chapter is structured as follows: after a short description of IPBES CF, we briefly outline the history of previous conceptual frameworks used in biodiversity and ecosystem service assessments. We then focus on the genesis of IPBES CF, to highlight the tensions and controversies that arose during its emergence, from 2012 to late 2013. Finally, we conclude by discussing what might be expected from IPBES CF in light of the controversies that punctuated its development, and reflect on its innovative nature. The results presented in this chapter are based on a range of different primary and secondary sources. They include the results of a research program on ecosystem services which analyzed the Millennium Ecosystem Assessment process; an analysis of comments received on the first draft proposal for an IPBES CF; a range of interviews (10) conducted with experts who participated in the development of IPBES CF; participant observations collected during two IPBES plenary sessions (IPBES-1, January 2013; IPBES-2, December 2013) and United Nations Environmental Programme (UNEP) documents disseminated for those plenaries.

IPBES conceptual framework in a nutshell

Following the official establishment of IPBES (Panama; April 2012), one of the first decisions made by IPBES delegations consisted in initiating the development of a common conceptual framework. The Panama meeting recommended

> to prepare a draft conceptual framework document informed by the review of assessments and drawing on existing conceptual frameworks. The draft will be made available to all Governments and stakeholders for online review through an open and transparent process, and all comments received will be compiled for consideration by a multidisciplinary and regionally balanced expert workshop that will be mandated to make a proposal for a conceptual framework for consideration by the Plenary at its first session.
>
> (UNEP, 2012b)

UNESCO, along with other organizations such as DIVERSITAS, was officially delegated with initiating the development of this CF. While reflections on a potential IPBES framework had started before, the process formally took place in the context of two major workshops organized between October 2012 and December 2013. IPBES CF was then officially adopted during the second IPBES Plenary session (IPBES-2, Antalya, December 2013). During this process, IPBES CF was described as:

> A concise summary in words or pictures of relationships between people and nature. In other words, conceptual frameworks depict key social and ecological components, and the relationships between these components. They provide common terminology and structure for the variables that are of interest in the system of interest, and propose assumptions about key relationships in the system.
>
> (UNEP, 2012a: 11)

As an outcome of a first expert workshop, a first conceptual framework was presented to IPBES delegations in January 2013 (Bonn, IPBES-1). An early document underlined the diverse functions that IPBES CF should fulfill, emphasizing that it:

> (1) help clarify and focus thinking about complex relationships, supporting communication across disciplines, knowledge systems and between science and policy, (2) may provide support to structure and prioritize work and (3) may also allow buy-in from a variety of stakeholders, by involving them in the development of the framework, and thus increase policy relevance.
>
> (Ibid.)

When it was presented during the second plenary in a more elaborate form, IPBES CF was characterized by three distinctive features regarding both the way it was developed and its content. These emphasized: (1) the quality of the process leading to its adoption, with the inclusion of different voices in order to ensure credibility and legitimacy; (2) the pivotal role of "institutions" and not "nature", hence highlighting the importance of governance aspects for managing biodiversity and ecosystem services; and (3) the integration of different knowledge-systems by means of a color code.[2] This color coding is the reason the IPBES CF was nicknamed a "Rosetta Stone" by some of its designers: "the conceptual framework can be thought of as a kind of Rosetta Stone that highlights commonalities between diverse value sets and seeks to facilitate crossdisciplinary and crosscultural understanding" (Díaz et al., 2015: 1).

However, this enthusiastic presentation of the IPBES conceptual framework, which highlights its inclusivity and innovative nature, should be balanced by a more detailed presentation of the elaboration process and the controversies it has generated, particularly around the ecosystem service concept. Before turning to these controversies, we present previous biodiversity and ecosystem service assessments which had also adopted a common conceptual framework. In fact, IPBES CF has its roots in the Millennium Ecosystem Assessment framework, which has largely contributed to the diffusion of the ecosystem service concept – even now included in the IPBES name.

Conceptual framework for biodiversity and ecosystem service assessments: the first steps

While many biodiversity and ecosystem service assessments have been carried out, two are of particular interest to contextualize the development of IPBES. First, the Global Biodiversity Assessment (GBA/1993–1995), which was conducted soon after the creation of the Convention on Biological Diversity (CBD 1992). The purpose of GBA was to provide a global assessment of existing knowledge in the biodiversity field. The second is the Millennium Ecosystem Assessment (MA/2001–2005) which was an essential milestone in the genesis of IPBES.

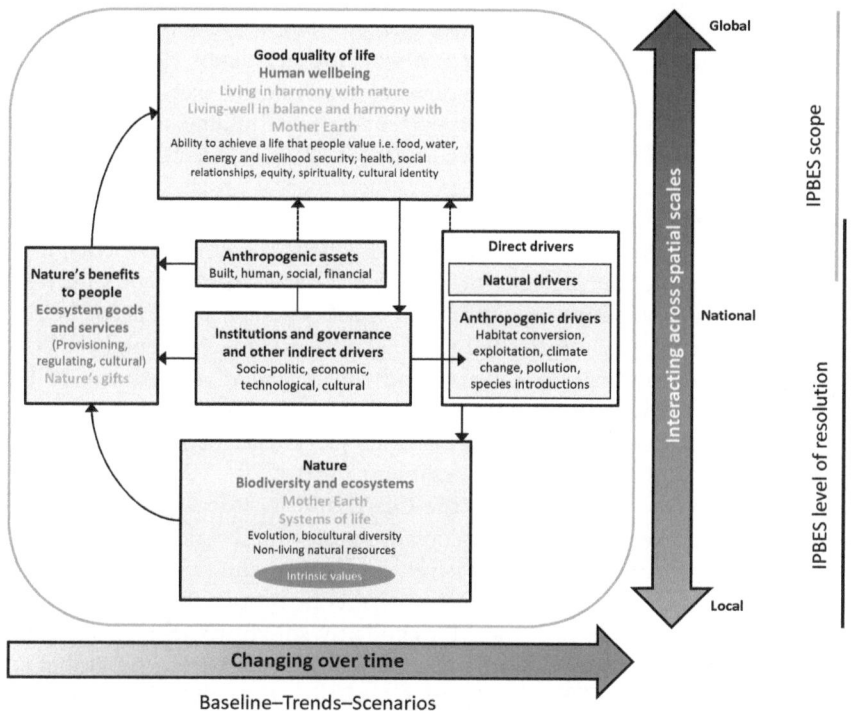

Figure 7.1 IPBES conceptual framework

GBA, which was convened by UNEP, mostly mobilized natural scientists and involved around 300 scientific experts. It was based exclusively on their contributions and all of the material was peer-reviewed only by scientists, and not by governments. The organizers initially planned to mobilize several disciplines, especially the various biological sciences (including population ecology, taxonomy, genetics and ecosystem ecology), socioeconomic sciences, and applied management sciences. However, the GBA structure and content suggests that social sciences were very little represented and that inter-disciplinary interactions did not really take place. Among the 13 chapters of the GBA report, only three fall outside of the biology field. Ten chapters approach biodiversity from the evolutionary biology and population ecology perspective and two chapters consider the relations between human activities and biodiversity and the value of biodiversity from an economic standpoint. Finally, one chapter looks at conservation measures through management sciences as applied to biodiversity (Heywood, 1995). Overall, the GBA was largely dominated by natural sciences. Particularly telling is the fact that the introduction of the GBA featured a simple conceptual diagram opposing human societies to biodiversity. Interactions between biodiversity and human societies were conceptualized through a sketchy circular conceptual framework – human societies having human influence on biodiversity and biodiversity having different values for human societies. At the center, underlying forces were mentioned but without any detail.

This first assessment, while recognized as scientifically excellent, was actually poorly received by governmental representatives – holding to its claim to be representative of independent science – without any particular political mandate. The GBA was not included in the intergovernmental process that started with the CBD (Cash and Clark, 2001). This means that while the GBA may have initiated some interactions and coordination efforts between scientists, it failed in its ambition to communicate the urgency of decision making to address the loss of biodiversity. Moreover, the GBA faced several criticisms for being too conservation-oriented – opposing human actions to environmental problems – and largely overlooking the poverty and development dimensions of biodiversity issues, therefore excluding critical issues of particular relevance for developing countries from the framing of its work (Biermann, 2002).

The Millennium Ecosystem Assessment (MA, 2001–2005) can be seen as largely constructed in reaction to the GBA, and its institutional design. The MA conveners were careful to associate several international organizations, as well as government representatives and representatives of diverse civil society sectors (businesses, NGOs), in its conception and implementation (Watson, 2005). In doing so, the MA explicitly sought to achieve "credibility, relevance, and legitimacy". While the GBA was focused on the global scale, the MA also innovated by conducting assessments at different scales, while paying attention to the multi-scalar dimension of biodiversity-related issues.[3]

The MA exercise was different from GBA by virtue of the organizers' intention to incorporate non-scientific knowledge, and also because they sought to expand the process to a variety of disciplines and encourage interactions between them. The MA mid-term assessment showed that ecologists, biologists, and economists were dominant in the process. However it did include a few sociologists, agronomists, physicists, geologists, hydrologists and anthropologists (Miller & Dublin, 2004). This inter-disciplinarity was facilitated by the mobilization of particular sub-disciplines. In particular, the predominance of functional ecology in the MA, as opposed to the GBA's population ecology focus, facilitated interactions between ecologists and economists. Beyond scientific knowledge, the MA also included expertise through the involvement of experts from the development sector (World Bank, FAO,[4] CGIAR,[5] etc.) and think tanks (IIED,[6] WRI,[7] etc.). Although one of the MA objectives was to incorporate "knowledge held by the private sector, practitioners, local communities, and indigenous peoples" (Millennium Ecosystem Assessment, 2005: v), only a handful of the 1,360 experts involved represented indigenous organizations and local knowledge, the private sector (Syngenta, BP), or environmental NGOs.

In trying to foster cross-disciplinary interactions, the MA participants formalized an explicit conceptual framework organized around the ecosystem service concept, i.e. "the benefits that humans gain from ecosystems" (Millennium Ecosystem Assessment, 2005). In promoting this concept, the

MA adopted a classification of different types of ecosystem services, including provisioning services, cultural services, regulating services and supporting services. In contrast to GBA, a major aspect of the MA is that via the ecosystem service concept it explicitly adopted an anthropocentric and utilitarian framing, therefore seeking to overcome the binary opposition between humans and nature: rather than opposing human actions to "nature", the MA highlights the connections between the diverse types of ecosystem services and the wide range of benefits they provide to societies. It is worth underscoring that the MA was explicitly intended to contribute to the Millennium Development Goals (2000–2015).

In this respect, the MA is ambitious in the sense that it proposes to analyze the interactions between biodiversity, ecosystem services, and human wellbeing, and the dynamics at the origin of the related changes as they emerge at different spatiotemporal scales (see below). By contrast to the GBA framework, the MA framework gives more details on the drivers affecting biodiversity loss, while making a distinction between indirect and direct drivers.

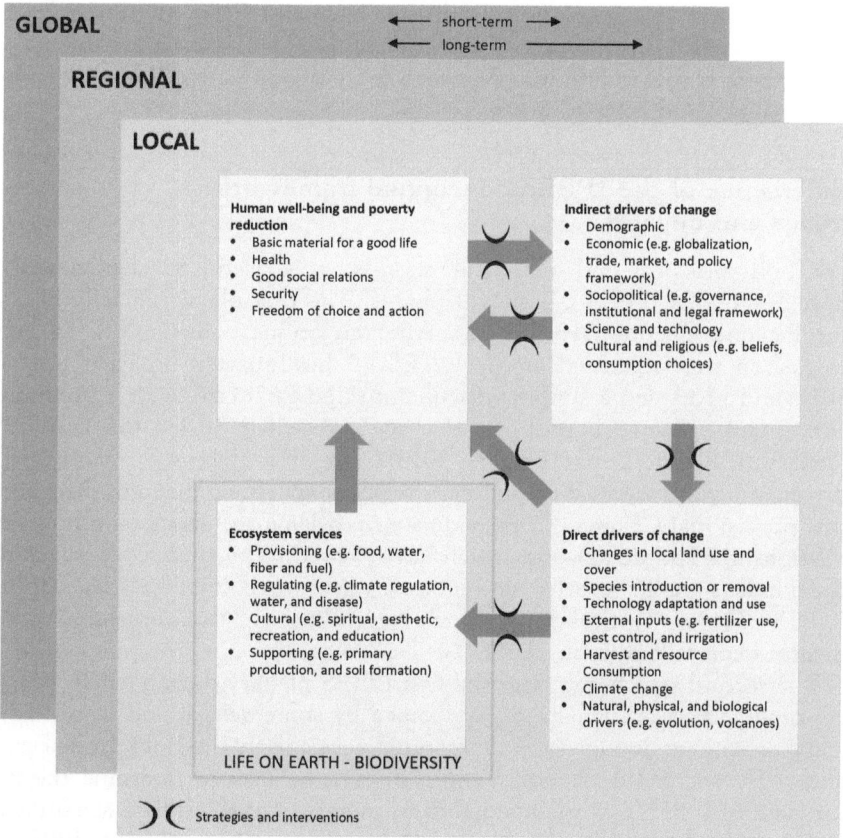

Figure 7.2 Millennium Ecosystem Assessment (MA) conceptual framework

After the assessment, the MA CF was criticized for analyzing ecosystems solely from the perspective of the services that they deliver to mankind. One of the main criticisms of MA relates specifically to the "ecosystem services" notion which is often perceived as too anthropocentric, and fails to do justice to the multiple ways of approaching human–nature relations (Peterson et al., 2010). For example, anthropologists have underlined that this concept is far from the social representation and aspirations of indigenous populations whose knowledge was largely overlooked in the analysis (Filer, 2009; Brosius, 2006). Science and Technologies Studies scholars also pointed out the performative effects that the ecosystem service approach may entail when it comes to protecting biodiversity in practice (Ernstson and Sörlin, 2013; Turnhout, Neves, and de Lijster, 2014).

To some extent, IPBES has its roots in the MA and several of the scientists and organizations previously involved in the MA are now mobilized in IPBES; these include in particular organizations such as UNEP, ICSU, DIVERSITAS (now Future Earth) and International Union for the Conservation of Nature (IUCN). Importantly, the MA Chair and Co-Chair (Robert Watson and Abdul Hamid Zakri) are now both Co-Chair and Chair of IPBES. Hence, although IPBES, in contrast to the MA, operates in intergovernmental settings, there is some continuity between these two initiatives in terms of both personalities and institutions.

Emergence of the IPBES conceptual framework: trends and debates

For both practical and analytical reasons, it is useful to characterize the development of IPBES CF in two phases. The first phase officially began at the Panama meeting (April 2012). After an online consultation, an expert workshop was organized at the UNESCO headquarters in Paris (October 2012) and produced a support document (UNEP, 2012a), with a preliminary conceptual framework proposal that was presented during the first IPBES Plenary (IPBES-1, Bonn, January 2013). This first stage was mainly expert driven and produced a first and very simple conceptual diagram. The aim of this provisional CF was "to provide a shared language and a common set of relationships and definitions, which have proved to be effective in supporting the kinds of collaborative and comparative work anticipated in IPBES" (UNEP, 2013a: 11), thus supporting a "coordinative" discourse among these diverse experts (Schmidt, 2008).

The second step began after the first IPBES plenary during which this first conceptual diagram, which was welcomed by some delegations, also triggered some discontent among a few delegations, particularly the Bolivian delegation, and some stakeholders. Following IPBES-1, the task to continue the development of IPBES CF was handed in to members of the IPBES scientific and technical body, i.e. the Multidisciplinary Expert Panel (MEP). IPBES CF underwent an endless number of metamorphoses. A second workshop was organized in Cape Town (August 2013), under the leadership of two of IPBES MEP members. The final framework was finally drafted during this event, which gathered government-nominated experts as well as IPBES MEP and Bureau members. It was officially accepted by IPBES delegations during IPBES-2 (Antalya, December 2013). This second stage gathered experts and governmental delegates, along with other stakeholders and highlighted the more "communicative dimension" of the conceptual framework (Schmidt, 2008). The IPBES framework can be seen not only as a tool to coordinate activities but also as an element through which IPBES attempts to produce meaning and coherence to obtain a broader impact on public opinion.

Specific controversies emerged in both phases. While during the first phase, these were mostly confined to disciplinary and interdisciplinary debates between diverse scientific communities, the second phase was more explicitly characterized by conflicting worldviews and politico-epistemic struggles regarding what approach should be adopted to address biodiversity issues.

Inception of the IPBES conceptual framework: coordinating expertise (from Panama 2012 to Bonn IPBES-1)

An expert meeting was organized in Tokyo in February 2012 and these experts recommended that IPBES CF be based on "a review of existing assessment frameworks, to develop a trans-disciplinary common conceptual framework, methods and approaches and maintain a dynamic catalogue of the relevant assessment landscape with the establishment of a state of art knowledge management system at the very beginning of IPBES" (UNU, 2012: 4). This international scientific workshop highlighted the early involvement of scientific experts in developing the future IPBES conceptual framework and the leading role of some scientific organizations[8] in the process, with the support of UNU and the governments of Japan and South Africa. Before that, informal discussions had also taken place amongst the members of DIVERSITAS. Participants in this expert workshop claimed to represent the "scientific community" and formulated "a strong recommendation for the development of a common conceptual framework to provide for consistent and coherent assessments at different scales and in different regions, developed in a transdisciplinary multi-knowledge way and addressing the needs of the different end users" (UNU, 2012: 3).

During the Panama meeting in April 2012, the Member States requested that the interim Secretariat

> prepare a draft conceptual framework document informed by the review of assessments and drawing on existing conceptual frameworks. The draft will be made available to all Governments and stakeholders for online review through an open and transparent process, and all comments received will be compiled for consideration by a multidisciplinary

and regionally balanced expert workshop that will be mandated to make a proposal for a conceptual framework for consideration by the Plenary at its first session.

(UNEP, 2012b: 25)

A first expert workshop was then organized in Paris: "Under the leadership of UNESCO, a small organizing committee was created" to prepare an informal experts workshop in Paris.[9] A list of invited experts was established by the organizing team, in association with the IPBES interim Secretariat and following the recommendations of the Panama meeting report: to be representative of a broad range of geographical locations (namely to have a North/South balance), of disciplines (ecologists, economists, anthropologists were invited). The Paris workshop was mainly a scientific experts' workshop: almost all of these experts had a PhD and most were still working either as scientists or in academia. Most of them also had experience with global change research programs such as DIVERSITAS and IHDP[10] (Borie and Hulme, 2015).

The main output of the workshop was an information document which identified 12 key messages and six considerations for the attention of delegates at the first IPBES Plenary meeting. This document also pooled an array of conceptual frameworks from other initiatives, such as MA, the United Kingdom National Ecosystem Assessment and The Economics of Ecosystems and Biodiversity (TEEB).

The first four key messages underlined the expectations raised by a conceptual framework. For the experts, conceptual frameworks are useful for encouraging agreement upon common definitions and facilitating the emergence of a shared vocabulary. In particular, participants in the Paris workshop suggested that: "Conceptual frameworks, if developed in an open and transparent process allowing the involvement of a broad set of stakeholders and knowledge holders can significantly increase policy relevance by addressing user needs as well as improving adaptation and learning" (Ibid.). The conceptual framework should not only guide assessment activities but also other IPBES functions. A major aspect is that the conceptual framework should allow multidisciplinary collaboration and enable communication across disciplines. In this respect, experts emphasized that the conceptual framework should also be inclusive of indigenous and local knowledge systems. However, as explained below, this dimension was not obvious in the first conceptual diagram.

The following seven key messages described the content of the proposed conceptual framework and the relations between its four building blocks: biodiversity and ecosystem functioning, ecosystem goods and services, human well-being and institutions and decisions as key indirect and direct drivers of all inter-linkages (see Figure 7.3).

The Paris experts' workshop highlighted the cultural dimension of human well-being and considered it as "multi-dimensional and dependent on access

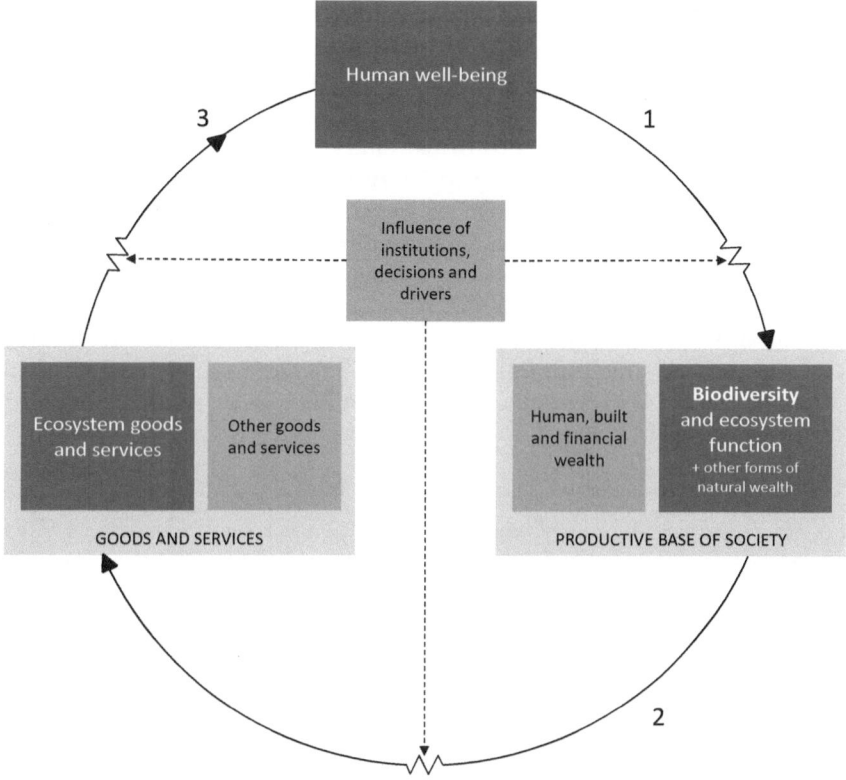

Figure 7.3 The "Paris diagram" presented during the first Plenary in Bonn, January
2013 (IPBES-1)

to and changes in bundles of goods and services and is context specific with
preferences for constituents of human well-being varying across individuals
and societies" (Ibid.: 6). Arrows were used to show interactions and causal
relations between these different building blocks. During the Paris workshop,
these arrows and their direction were debated endlessly. For instance, some
participants wanted to introduce a double-arrow between human well-being
and biodiversity, arguing that biodiversity has intrinsic value and, consequently, benefits directly to humans without necessitating the mediation of
ecosystem goods and services. The natural capital concept was also discussed
and avoided in the final draft, which refers instead to assets and wealth,
words that were perceived as being more "neutral".[11] In the Paris diagram,
"human, built and financial wealth" and "biodiversity and ecosystem function" form the "productive base of society". Many discussions also took
place regarding the ways in which the classification of ecosystem services
promoted by MA should be used: participants often stressed the fact that
some categories of ecosystem services (e.g. supporting services) were not
easy to work with as analytical categories.

The introduction of institutions, as framed by but also framing human decisions, was presented as a significant change in comparison with the MA. For the Paris experts

> institutions encompass all formal and informal interactions among stakeholders and structures that determine how decisions are taken and implemented, how power is exercised, and how responsibilities are distributed. Institutions also determine to various degrees the access to, allocation and distribution of the various forms of wealth and the benefits they provide through goods and services. They can be organized along a continuum of temporal and geographical scales spanning from global institutions to small groups and individuals, influencing and influenced by socio-economic-cultural contexts, including values, traditions, customs, norms and fads.
>
> (UNEP, 2012a: 22)

This institutional approach was clearly influenced by the work of Elinor Ostrom, whose ideas and concepts were already influential in the MA. The need to integrate "institutions" is based on the willingness to underline the fact that governance settings, including regulations and institutional dynamics, have to be taken into consideration to address biodiversity and ecosystem service issues. This reflects the idea that "biodiversity issues" need to be addressed by including more than natural science knowledge – economists and social scientists have a key contribution to make by documenting these aspects.

Enlarging the debate: a contested framework (from IPBES-1 Bonn 2013 to Antalya 2013)

Following the Paris workshop, the resulting conceptual framework was posted online on the IPBES website before IPBES-1 and delegations, as well as members of civil society and other stakeholders, were given the possibility to express their views on it. Overall, 27 written comments were received and made available online; these included 12 reactions by states[12] and 12 by stakeholders.[13] This process showcased several positions: while many delegations did not react to the framework, some welcomed it (e.g. India) or suggested that it was a good starting point for further discussions (e.g. Germany). However, it was also strongly contested by other delegations, in particular by the representative of Bolivia, supported by other South American delegations from ALBA,[14] and by several stakeholders.

As underlined by some participants at the Paris workshop:

> There were comments from a whole range of people, individuals, governments, on the document. And it became very clear that some people were very uncomfortable with it, particularly because it looks like that there is not value for nature itself. [The Paris diagram] was a completely

utilitarian view and apparently it was also a very Western view of the world, which is probably true...although we did have a wide range of people at that workshop.[...]There was this very strong argument against it from the Bolivian delegation but also a number of other South American Nations and Parliaments that said that this just did not work for them. They wanted Mother Nature, they came up with their own one, with Mother Nature in the middle.[15]

In Bonn, the MEP experts and Bureau members were nominated. The Plenary did not reach a consensus on the draft conceptual framework and mandated the MEP to organize "a multidisciplinary and regionally balanced expert and stakeholder workshop, among other actions, to provide inputs to the development of a draft conceptual framework for the Platform[...]for adoption by the Plenary at its second session" (UNEP, 2013c: 28). In fact, few debates on IPBES CF explicitly took place during the Plenary. Some delegations expressed their views on it (e.g. Bolivia) but most of the detailed comments were expressed through written communications.

The Paris diagram was perceived as too utilitarian by a wide range of actors. Their comments emphasized that it did not adequately reflect the plurality of values that could be attributed to biodiversity. For the Helmholtz-Centre for Environmental Research – UFZ (Germany): "The current draft seems to be biased towards an economic perspective (...). There is a need to address cultural and ethical considerations in more details". Similar comments were expressed by the NGO BirdLife International: "The emphasis is on utilitarian values throughout. This in turn does not reflect the differing emphasis put on the different values of biodiversity by different people and different cultures" and by the UK representative: "There is a strong emphasis on ES and a utilitarian perspective and we believe that this should be further complemented by the inclusion of other values of biodiversity". FAO also insisted on this aspect, stating that: "The conceptual framework insists mainly on the economic dimension of biodiversity. The conceptual framework might be rebalanced to take into account social and cultural dimensions."

But the strongest contestation came from representatives of the Plurinational State of Bolivia. From this perspective, the initial IPBES CF was perceived as: (1) representative of a "Western modern society view" focusing on the use and exploitation of nature by individuals solely interested in their own well-being; (2) too centered on an economic understanding of nature; and (3) reducing ecosystem functions to the production of goods and services and therefore giving an important role to markets to achieve this transformation. Moreover, for the Bolivian delegation, the conceptual framework did not consider the distinctive views of people and societies, and excluded indigenous and local knowledge. It was perceived as "a tool for instrumenting the development of the privatization and commodification of nature" (Plurinational State of Bolivia, 2013: 3). The Bolivian delegation formulated an alternative proposal: "the Living-Well in balance and harmony with Mother

Earth (LME)". This alternative way of approaching human–nature relations was based on ethical values, underpinned by an holistic worldview, and it attributed rights to nature. It also included references from Ostrom's work on polycentric governance and on the nature of goods.[16] This proposal was strongly defended by the Bolivian delegation during the IPBES-1 meeting, with other supporting voices from Guatemala and more discrete support from several South American countries. Conversely, other countries, like Canada and the USA, advocated a more economic perspective, pushing for integration of the natural capital concept.[17]

Following IPBES-1, some informal contacts where established between the Bolivian representative and some of the experts previously involved in the Paris meeting. This informal discussion reduced mutual distrust between the Bolivian position, already expressed in other fora such as CBD, and scientists advocating the – now hegemonic – ecosystem service approach. During this second phase, productive interactions between actors not used to working together were thus encouraged. It is worth underlining here that UNESCO, as a major actor in this process, asked for a desk study on the Bolivian laws to better understand the criticism from this country and played a facilitating role to reach a compromise (Limache de la Fuente & Arico, 2015).

Building a consensus? A color code for the Rosetta Stone

The Bonn Plenary mandated the MEP to organize two workshops, one on different knowledge systems and another on the conceptual framework. The Tokyo meeting (June 2013) mainly focused on identifying procedures and possible approaches to consider indigenous and local knowledge in IPBES. It also drew up a few recommendations for the conceptual framework, advocating broadening the consultation process: "to reflect the multiple representations of relations between social and ecological spheres both in terms of science-based conceptual frameworks as well as diverse indigenous and local world-views" (UNESCO, 2013: 18). The Tokyo meeting suggested that "IPBES should critically evaluate the appropriateness of the ecosystem services framework and its current priority setting tools for equitable allocation of resources to restore indigenous and local community wellbeing" (Ibid.). However, very few participants of the Tokyo meeting also participated in the second workshop devoted to the IPBES CF.

During the intersessional process, a second major workshop devoted to the development of IPBES CF was organized in Cape Town (South Africa) in August 2013. This workshop gathered 28 experts selected by the MEP,[18] 29 IPBES MEP and Bureau members and nine representatives from Multilateral Environment Agreements, scientific subsidiary bodies, and UN organizations. Two MEP members, i.e. Sandra Diaz (ecologist, Argentina) and Sebsebe Demissew (botanist, Ethiopia) co-chaired the workshop. This two-day meeting provided a space for heated debates on the conceptual framework, with again some critics of the ecosystem service concept and its potentially utilitarian vision. However, the development of IPBES CF was much more than just about ecosystem services – the whole process was punctuated by numerous debates which were not always focused on this notion. Following numerous exchanges between dissenting views, a compromise was reached during the Cape Town workshop: "It was that very powerful set of interventions from the Bolivians that really re-framed it and many discussions took place with South American delegates involved in IPBES" (ecologist, participant in the Paris and Cape Town workshops).

In order to overcome the divide between Mother Earth and ecosystem services, both views were conflated on the same conceptual diagram by means of a color code: "Text in green denotes the concepts of science; and text in blue denotes those of other knowledge systems" (UNEP 2014). The IPBES CF has therefore been compared to a Rosetta Stone, a metaphor which suggests that this device has the ability to render commensurable two different ways of framing biodiversity. Indeed, the final conceptual diagram marks the recognition of both, the ecosystem service view and the Mother Earth view. Yet, in this framework the ecosystem services view is equated with the scientific view; while all other knowledge systems, thus depicted as non-scientific, become equated with Mother Earth. Moreover, "institutions" and not "nature" appear in the center – the framing is therefore clearly anthropocentric, and attributes a prominent role to governments. From a critical standpoint, it fails to actually recognize that different worldviews cannot fit in the same boxes.

The key facilitators of the global process over these two years of consultation and negotiation also acknowledged that this framework is the outcome of a compromise: "the CF is the result of political negotiations, but it goes beyond that.[…]The consultative construction process that converged in the model adopted by the IPBES Second Plenary was rich in discussion and conflict on epistemological and methodological, as well as political, grounds" (Díaz et al., 2015: 4).

Conclusion

During the process leading to the adoption of the IPBES conceptual framework, the notion of ecosystem services was contested and different viewpoints on the ecosystem service approach were expressed. While supported as an important epistemic notion to further address biodiversity issues by numerous scientists and delegations, it was also contested by others – most vehemently by the Bolivian delegation. Driven by the willingness to adopt a consensual overarching diagram, IPBES ended up adopting an innovative, yet intriguing, conceptual framework. The use of the Rosetta Stone metaphor highlights the metaphorical character of the communicative discourse around IPBES CF.

On the one hand, the recognition of Mother Earth on the diagram can be seen as a way to overcome conflicts and neutralize the critique targeting ecosystem services. The conceptual framework thus becomes a stabilizing device and ensures that IPBES will be able to continue its work, thus apparently nullifying the critique. On the other hand, as emphasized above, beyond Bolivia, many delegations and participants questioned the economic approach prevailing in the Paris diagram and stressed the importance of including other values of biodiversity. In this respect, while advocating Mother Earth, South American delegations may have opened a space for other approaches and ways of knowing biodiversity, even in the Western world. For some people, significant progress has been made since the MA conceptual framework. Institutional aspects are increasingly considered as being relevant to addressing biodiversity-related issues, and the diversity of knowledge, from Western science to other knowledge systems, has been taken into consideration through a color code that renders them both visually equivalent. But IPBES CF remains clearly embedded in the MA perspective, with the ecosystem service concept having a key role.

Importantly, IPBES CF is meant to be used in practice. It is supposed to support the implementation of all four functions of the IPBES work program and help to ensure coherence and coordination among them (UNEP, 2013b). IPBES CF is explicitly considered as a device which, while providing a vision for IPBES, should also be used in the implementation of its operations. In this respect, instead of being a "poster to hang on the wall" (or a nice image), the ambition of IPBES is to "operationalize" this Rosetta Stone, despite the ambiguities outlined above. In contrast to GBA and MA, this framework recognizes different ways of knowing, consistent with the ambition of IPBES to be a knowledge platform and not a "science" platform.

However, many doubts remain regarding what can realistically be expected from IPBES in terms of considering diverse knowledge while being sensitive to a variety of local situations and working at multiple scales. In particular, some have suggested that considering local knowledge at a global scale may be too ambitious for IPBES and that the organization may become a "tower of Babel of scale and cultures" (Soberón and Peterson, 2015). Moreover, the findings of the analysis of the composition of the first and second MEP suggest that, despite the disciplinary and gender balance principles advocated by IPBES, so far this expert body remains dominated by natural scientists with academic credentials. The consideration of indigenous and local knowledge remains controversial (Montana and Borie 2015). Will IPBES CF really help integrate different knowledge systems or, from a more communicative standpoint, more simply superimpose different visions to send signs of goodwill? Future research is needed to determine whether IPBES will be able to integrate these different knowledges in practice, and with what consequences for the governance of biodiversity.

Notes

1 See "IPBES mission" in http://www.ipbes.net/about-us.
2 In the IPBES conceptual framework, "text in green denotes the concepts of science and text in blue those of other knowledge systems". Text in black is supposed to be shared by all the worldviews.

3 From the regional Southern African Millennium Ecosystem Assessment to the very local focus like the Stockholm Urban Assessment.
4 Food and Agriculture Organization of the United Nations.
5 Consortium of International Agricultural Research Centers.
6 International Institute for Environment and Development.
7 World Resources Institute.
8 Mainly the International Council of Science Unions (ICSU) represented by two international programs: the International Program of Biodiversity Science (DIVERSITAS) and the International Human Dimensions Programme on Global Environmental Change (IHDP)
9 It includes Salvatore Arico (ecologist, UNESCO), Neville Ash (ecologist, UNEP), Eduardo Brondizio (anthropologist, Indiana University, USA), Anne Larigauderie (ecologist, Executive Director of the International Council for Science), Georgina Mace (ecologist at University College London, Head of DIVERSITAS), Kazuhiko Takeuchi (geographer, Vice-Rector of the United Nations University, Tokyo) and Pierre Commenville (ecologist, IUCN).
10 International Human Dimensions Programme on Global Environmental Change.
11 Interview with participant of the Paris workshop (social scientist based in Europe).
12 Bolivia, Canada, UK, France, Germany, India, Japan, New Zealand, Norway, The Netherlands, USA.
13 Among other, Biogenesis, ICSU, Applied Environmental Research Foundation (AERF), World Conservation Monitoring Centre (UNEP-WCMC), United Nations University Institute of Advanced Studies (UNU-IAS), Birdlife International, FAO, IUCN, University of Melbourne, Wildlife Conservation Society (WCS).
14 *Alianza Bolivariana para los Pueblos de Nuestra América*.
15 Interview with an IPBES contributing expert (ecologist based in Europe).
16 The main Bolivian expert in IPBES has a PhD supervised by E. Ostrom on resources management – this was probably a facilitating factor to build a compromise for the IPBES conceptual framework.
17 Interview with a participant of IPBES-1 (ecologist, international organization).
18 Three of those participants were selected to share the results of other expert workshops on Indigenous and Local Knowledge Systems which took place in Tokyo, Japan, in June 2013 to provide the conceptual framework workshop with inputs related to knowledge systems.

References

Biermann, F. 2002. Institutions for Scientific Advice: Global Environmental Assessments and Their Influence in Developing Countries. *Global Governance*, 8, 195–219.
Borie, M. & Hulme, M. 2015. Framing Global Biodiversity: IPBES between Mother Earth and Ecosystem Services. *Environmental Science and Policy*, 54, 487–496.
Brosius, P. 2006. What Counts as Local Knowledge in Global Environmental Assessments and Conventions? *In*: Reid, W. V., Berkes, F., Wilbanks, T. & Capistrano, D. (eds.) *Bridging Scales and Knowledge Systems: Concepts and Applications in Ecosystem Assessment/Millennium Ecosystem Assessment*. Washington: Island Press.
Cash, D. W. & Clark, W. C. 2001. From Science to Policy: Assessing the Assessment Process. *Faculty Research Working Papers Series*. Cambridge, MA: John F. Kennedy School of Government, Harvard University.
Clark, W. C., Mitchell, R. B. & Cash, D. W. 2006. Evaluating the Influence of Global Environmental Assessments. In: Mitchell, R. B., Clark, W. C., Cash, D. W.

& Dickson, N. M. (eds.) *Global Environmental Assessments: Information and Influence*. Cambridge, MA: MIT.

Diaz, S., Demissew, S., Joly, C., Lonsdale, W. M. & Larigauderie, A. 2015. A Rosetta Stone for Nature's Benefits to People. *PLOS Biology*, 13(1), 1–8.

Ernstson, H. & Sörlin, S. 2013. Ecosystem Services as Technology of Globalization: On Articulating Values in Urban Nature. *Ecological Economics*, 86, 274–284.

Farrell, A., Vandeveer, S. D. & Jäger, J. 2001. Environmental Assessments: Four Under-appreciated Elements of Design. *Global Environmental Change*, 11, 311–333.

Filer, C. 2009. A Bridge Too Far: The Knowledge Problem in the Millennium Assessment. *In*: Carrier, J. G. & West, P. (eds.) *Virtualism, Governance and Practice. Vision and Execution in Environmental Conservation*. New York/Oxford: Berghahn Books.

Gieryn, T. F. 1983. Boundary-Work and the Demarcation of Science from Non-Science: Strains and Interests in Professional Ideologies of Scientists. *American Sociological Review*, 48, 781–795.

Guston, D. H. 2001. Boundary Organizations in Environmental Policy and Science: An Introduction. *Science, Technology & Human Values*, 26, 399–408.

Heywood, V. H. 1995. *The Global Biodiversity Assessment*. United Nations Environment Programme. Cambridge: Cambridge University Press.

Hilgartner, S. 2004. The Credibility of Science on Stage. *Social Studies of Science*, 34, 443–452.

Koetz, T., Farrell, K. & Bridgewater, P. 2012. Building Better Science-policy Interfaces for International Environmental Governance: Assessing Potential within the Intergovernmental Platform for Biodiversity and Ecosystem Services. *International Environmental Agreements: Politics, Law and Economics*, 12, 1–21.

Limache de la Fuente, D. & Arico, S. 2015. La ley de derechos de la Tierra Madre para el mundo. *La Migrana. Revista de Analisis Politico*, 4.

Millennium Ecosystem Assessment 2005. *Ecosystems and Human Well-being: Synthesis*. Washington, DC: Island Press.

Miller, C. A. & Dublin, H. 2004. *Millennium Ecosystem Assessment, UNEP/GEF/ UNFIP PROJECT – MP/FP/1010–1004, Mid-term Evaluation*. Nairobi: UNEP/GEF.

Mitchell, R. B., Clark, W. C., Cash, D. W. & Dickson, N. M. (eds.) 2006. *Global Environmental Assessments: Information and Influence*. Cambridge, MA: MIT.

Montana, J. & Borie, M. 2015. IPBES and Biodiversity Expertise: Regional, Gender, and Disciplinary Balance in the Composition of the Interim and 2015 Multidisciplinary Expert Panel. *Conservation Letters*, 1–5.

Peterson, M. J., Hall, D. M., Feldpausch-Parker, A. M. & Peterson, T. R. 2010. Obscuring Ecosystem Function with Application of the Ecosystem Services Concept. *Conservation Biology*, 24, 113–119.

Plurinational State of Bolivia 2013. *Conceptual Framework for the Intergovernmental Science-Policy Platform on Biodiversity and Ecosystem Services*.

Schmidt, V. A. 2008. Discursive Institutionalism: The Explanatory Power of Ideas and Discourse. *Annual Review of Political Science*, 11, 303–326.

Soberón, J. & Peterson, A. T. 2015. Biodiversity Governance: A Tower of Babel of Scales and Cultures. *PLoS Biology*, 13, e1002108.

Turnhout, E., Neves, K. & De Lijster, E. 2014. "Measurementality" in Biodiversity Governance: Knowledge, Transparency, and the Intergovernmental Science – Policy Platform on Biodiversity and Ecosystem Services (IPBES). *Environment and Planning A*, 46, 581–597.

UNEP 2012a. Outcome of an Informal Expert Workshop on Main Issues Relating to the Development of a Conceptual Framework for the Intergovernmental Science-Policy Platform on Biodiversity and Ecosystem Services (IPBES/1/INF/9). Paris: United Nations Environment Programme (UNEP).

UNEP 2012b. Report of the Second Session of the Plenary Meeting to Determine Modalities and Institutional Arrangements for an Intergovernmental Science-policy Platform on Biodiversity and Ecosystem Services (UNEP/IPBES.MI/2/9). Panama City: United Nations Environment Programme (UNEP).

UNEP 2013a. Outcome of an Informal Expert Workshop on Main Issues Relating to the Development of a Conceptual Framework for the Intergovernmental Science-Policy Platform on Biodiversity and Ecosystem Services (IPBES/1/INF/9). Bonn, Germany: United Nations Environment Programme (UNEP).

UNEP 2013b. Recommended Conceptual Framework of the Intergovernmental Science-Policy Platform on Biodiversity and Ecosystem Services (IPBES/2/4). United Nations Environment Programme (UNEP).

UNEP 2013c. Report of the First Session of the Plenary of the Intergovernmental Science-Policy Platform on Biodiversity and Ecosystem Services (IPBES/1/12). Bonn, Germany: United Nations Environment Programme (UNEP).

UNEP 2014. Report of the Second Session of the Plenary of the Intergovernmental Science-Policy Platform on Biodiversity and Ecosystem Services (IPBES/2/17). Antalya, Turkey: United Nations Environment Programme (UNEP).

UNESCO 2013. The Contribution of Indigenous and Local Knowledge Systems to IPBES: Building Synergies with Science. UNESCO, UNU, IPBES.

UNU (United Nations University) 2012. Summary of Results from the Second International Science Workshop on Assessments for IPBES. 27–29 February 2012. Tokyo: United Nations University.

Van den Hove, S. 2007. A Rationale for Science-policy Interfaces. *Futures*, 39, 807–826.

Watson, R. T. 2005. Turning Science into Policy: Challenges and Experiences from the Science-policy Interface. *Philosophical Transactions of the Royal Society B: Biological Sciences*, 360. Available at: http://rstb.royalsocietypublishing.org/content/360/1454/471.

8 Building process, effectiveness and limits of an IPBES stakeholder group

Marie Hrabanski, Mohamed Oubenal and Denis Pesche

Introduction

The objective of the International Platform for Biodiversity and Ecosystem Services (IPBES) is to strengthen the science-policy interface for the conservation and sustainable use of biodiversity, long-term human wellbeing and sustainable development. IPBES was launched in 2012 after a long process that started in 2008. Stakeholders participated in the inception of the platform and are also expected to play a key role in forthcoming IPBES assessments and reports. In order to encourage stakeholder participation and self-organization, an institutional process is under way, coordinated by the IPBES Secretariat in collaboration with two main organizations: the International Union for Conservation of Nature (IUCN – see Box 8.1) and the International Council for Science (ICSU – see Box 8.2). The "stakeholder days" just before conference plenaries and other informal meetings have led to the establishment of a stakeholder engagement strategy (SES) which "has been identified as an important element for the relevance, effectiveness, credibility and overall success of the Platform" (IPBES/3/16, 2014).

According to the IPBES Secretariat, the establishment of a stakeholder engagement strategy is an institutional innovation that is not yet found in other science-policy interfaces, notably the Intergovernmental Panel on Climate Change (IPCC). The identification of IPBES stakeholders is a complex task as

> there is a great diversity of individuals, organizations, networks, programmes and constituencies working across different sectors and scales (local, national, sub-regional, regional and global) that need to be considered. The complexity of these tasks is increased by the levels in which stakeholders operate (local, national, sub-regional, regional, global). Other parameters include the diversity of disciplines (natural, social and economic sciences), the different types of knowledge (traditional, local and indigenous) and the different sectors (industry, health, food, energy, etc.). The cultural differences, language barriers, differing

stakeholder interests and different mandates and governance arrangements represent additional factors to take into consideration.

(IPBES, 2014: 6)

Despite this diversity, the IPBES Secretariat makes it possible to take the plurality of knowledge and actors into account and institutionalize that plurality within a single dynamic process with SES.

Box 8.1 International Union for Conservation of Nature (IUCN)

IUCN is the world's oldest and largest global environmental organization, with more than 1,200 governments and NGO members. IUCN's work is supported by over 1,000 staff in 45 offices and hundreds of partners in public, NGO and private sectors around the world (source: IUCN website, 2015). IUCN is a hybrid NGO: while IUCN is an international conservationist network, it is also an organization that brings together international organizations, states and national and international NGOs, and plays a role similar to a United Nations' organization yet without its statutes.

Box 8.2 International Council for Science (ICSU)

ICSU is a non-governmental organization with a global membership of national scientific bodies (122 members, representing 142 countries) and International Scientific Unions (31 members). ICSU's strategic vision is for a world where science is used for the benefit of all, excellence in science is valued and scientific knowledge is effectively linked to policy making (source: ICSU website, 2015).

A major challenge in global governance research is to gain further insight into the dynamics of power and influence among this diversity of stakeholders. For instance, the material and organizational assets of large firms give them strong influence capacity, while NGOs rely mainly on information and expertise, activism, and a claim of legitimacy (Haufler, 2009). Betsill and Corell think that the influence of NGOs on global environmental governance is based on their ability to affect the behavior of other actors through the intentional transmission of information (Betsill and Corell, 2008; Corell and Betsill, 2001). This chapter takes a look at the construction process of the IPBES stakeholder group and its functioning. Among the diverse range of stakeholders, how can heterogeneity be analyzed in terms of capacity and influence? To address heterogeneity, how do the Secretariat and its mandated organizations manage the construction of a coherent stakeholder strategy? Is the stakeholder group offering the same advantages for each stakeholder?

We put forward two main hypotheses. The first is that different stakeholders have different capacities (organizational, financial, symbolic, etc.) to influence and participate in IPBES. The second is that the Secretariat and its mandated organizations try to manage this heterogeneity by boosting stakeholders' awareness on the "good rules of conduct" to avoid conflict and build a common stakeholder position. This restrictive management, however, could lead to diversified strategies.

In the first part, we analyze how the stakeholder concept is used by IPBES, showing that this concept brings together perfectly heterogeneous organizations through their purpose, interests and means, and that the roles assigned to stakeholders are also diversified. In the second part, we analyze the dynamics of power and influence within this diversity of involved stakeholders. The third part deals with the way the IPBES Secretariat organizes the stakeholder days, manages and controls these meetings. Finally, in the third part, we propose a discussion on the inequality between stakeholders and the consequences for IPBES.

Stakeholder engagement strategy: definitions and challenges

In IPBES, the stakeholder concept brings together perfectly heterogeneous organizations. This very broad scope could be explained by the fact that the stakeholder concept and the idea of stakeholder involvement or engagement strategy were directly imported from firm management theory concerning corporate responsibility (Partridge et al., 2004). For business organizations, stakeholders can be thought of as any group or individual who can affect or be affected by a corporation or its activities (Freeman et al., 2010). Consequently, the identification and eligibility criteria are still not constant in formulating the stakeholder engagement strategy developed by IPBES.

From 2008 to 2012, the term "stakeholder" was discussed in IPBES meetings to develop a broad definition of stakeholders. Indeed, right from the first meeting that was held to consider the setting up of IPBES in November 2008, the issue of the role played by non-governmental stakeholders was a focus of discussions: "there was broad agreement that the platform should be intergovernmental, but there was a range of views on how to involve other stakeholders" (ENB, 2008: 9). This general option followed on from the Millennium Ecosystem Assessment (MA) which was governed by a multi-stakeholder board that included representatives of international institutions, governments, business, NGOs, and indigenous groups – this option was considered as positive (Bernstein, 2010). More globally, the UN recognizes that the involvement of a diverse range of actors, including those from civil society and the private sector, as well as local authorities and parliamentarians, is essential for effective action on global priorities (United Nations, 2004). Stakeholder involvement also serves to democratize global environmental governance (Bäckstrand, 2006). An online consultation was then undertaken by the United Nations Environment Programme (UNEP), ICSU and IUCN between August and September 2009 (UNEP, 2009). The survey showed that IPBES stakeholders were mainly "scientific and technical information providers" including 59% academic establishments, 19% NGOs and a more incongruous set of 22% made up of representatives of international organizations, the private sector, consultants and students. Europe (34%) and North America (22%) were in the majority, ahead of Asia (20%), Africa and Latin America/ the Caribbean (12%). At the meeting in Busan (UNEP, 2010), the discussions focused on the participation of "relevant stakeholders" as observers, examining different options for their inclusion in IPBES governance (Bernstein, 2010). This discussion took place in a context where an IPCC external review highlighted the need for this science-policy interface to include non-governmental stakeholders (InterAcademy Council, 2010).

At the first plenary session in Bonn in January 2013, IUCN and ICSU were given a mandate to coordinate drafting of a tentative stakeholder engagement strategy with a view to supporting the IPBES work program. Moreover, as shown by Kowalski and Jenkins who used social network analysis, non-governmental organizations (NGOs) such as IUCN and ICSU often act as bridging organizations within the natural resource governance arena (Kowalski and Jenkins, 2015). Considerable technical, financial, and personnel resources enable these two NGOs to effectively link actors across institutional and spatial boundaries (Mitchell, 2010).

Stakeholders have a pluralistic role in the stakeholder engagement strategy (SES). First, it is a fairly utilitarian vision of stakeholders, which subdivides stakeholders into two categories: contributors and end users (IPBES/3/16, 2014). The contributors consist of scientists, knowledge holders and practitioners. These terms are intended to include a broad range of knowledge holders, knowledge producers and knowledge managers. The terms also include holders of indigenous and local knowledge, applied researchers working for NGOs, scientific and knowledge networks, citizens involved in monitoring and citizen science, policy think tanks, universities, etc. (IPBES/3/16, 2014). The users are policy makers, including government officials of different levels and their advisors from the Platform member states, officials in UN specialized agencies and multilateral environmental agreements, intergovernmental and non-governmental organizations, development banks, local governments and local policy makers (IPBES/3/16, 2014). Second, the SES is also thought of as being linked to the IPBES communication and awareness strategy and stakeholders are presented as actors who can participate in the development of IPBES strategic partnerships with other multilateral environmental agreements such as the Convention on Biological Diversity (CBD). Lastly, they are also considered as facilitators who enable the identification of resource persons of use for IPBES activities (UNEP, 2013). In 2014, this strategy was fine-tuned and more clearly dissociated from the IPBES communication strategy:

> The overall purpose of stakeholder engagement is to support the implementation of the Platform's work programme and its specific

deliverables for the period 2014–2018 in a participatory, inclusive and transparent manner. The Platform will depend on expert individual contributions for its assessments. In addition, the other three functions of the Platform also require input from and participation by a diverse mix of stakeholders. Accordingly, the strategy for engaging with stakeholders is a key element of the efforts to mobilize support for the implementation of the Platform's work programme for the 2014–2018 period.

(IPBES, 2014: 2)

However, the rules to define the relevant stakeholders for SES are not totally defined. The IPBES Secretariat is aware of the diversity of potential stakeholders in the process and suggests categorizing them. On the IPBES website, the stakeholder page suggests six categories: governments and multilateral environmental agreements, NGOs, indigenous peoples and local communities, private sector, scientific community, and UN agencies. As the goal of the stakeholder engagement strategy is to support the work program, IPBES seeks to create convergent dynamics of this heterogeneous group of stakeholders around its objectives and deliverables. In this way, the rules for including stakeholders in assessments were clarified in the revised strategy document (IPBES, 2014). The question of selecting stakeholders was clearly raised in order to limit their number and guarantee their "relevance" for the IPBES process. Before each plenary session, a selection process is carried out directly, by way of a solicitation and accreditation rationale, and indirectly through the involvement dynamics of stakeholders who decide to take part in the process, by attending several meetings or, on the contrary, to no longer take part in them.

This utilitarian vision of stakeholder engagement could also help in shaping the political process at IPBES (ENB, 2015). A survey conducted by ICSU and IUCN identified potential incentives for organizations to engage as stakeholders in the work of the Platform, while highlighting the opportunity to influence this work (IPBES, 2014).

Furthermore, discussions on relevant stakeholders and the composition of the IPBES stakeholder group are crucial for governance of the stakeholder engagement strategy. SES governance was discussed and most nations were in favor of a strategy that was all inclusive, while leaving management of the stakeholder forum to the stakeholders. In contrast, few nations asked for restricted stakeholder membership with control by the Secretariat and governments (ENB, 2015). Finally, the Secretariat, under the supervision of the Bureau and the Plenary and working in collaboration with an inclusive, open-ended forum of stakeholders, will implement and operationalize SES. However, the eligibility criteria of relevant stakeholders are still controversial, which is why some delegates suggest that the stakeholder definition be revised as the current definition incorrectly included governments, while some others proposed to limit the definition to types of contributors and end users (ENB, 2015). Indeed, the stakeholder group composition is crucial because of the diversity and capacities and power of influence of the organizations involved.

Heterogeneity of IPBES stakeholders

Even if, according to IPBES, stakeholders refer to governmental actors and non-state actors, concretely the IPBES stakeholder group mainly involves non-state actors (NSA). NSAs are recognized as key players in international relations, and their role has grown in importance over the past 20 years (Bäckstrand, 2006; Bernstein and Cashore, 2007; Pattberg and Stripple, 2008) They strive to influence the development of regime structures, and establish policies (Arts, 2005; Betsill and Corell, 2008; Keck and Sikkink, 1998). They also participate in the application of mechanisms and their monitoring. This complex, multifaceted and evolving group of NSAs involved in global environmental governance includes transnational corporations (Clapp, 2005), indigenous people, environmental non-governmental organizations (ENGO) and other experts or civil society groups (Alcock, 2008; Conca, 1995). The following analysis allowed us to deconstruct the sample group of IPBES stakeholders. These categories reveal the dynamics of power and influence within this diversity of stakeholders.

Attribute-based construction of IPBES stakeholder profiles – organization type

The first stage consisted of identifying stakeholders involved in the "IPBES stakeholder engagement process". The emergence of the IPBES began with the first meeting in November 2008 in Putrajaya (Malaysia). Three more intergovernmental meetings marked the process (Busan, 2010; Nairobi, 2011; and Panama, 2012) before the January 2013 first Plenary session in Bonn, Germany. A total of six meetings, which could be qualified as "global", were organized subsequent to the emergence of IPBES. Different types of stakeholders participated in each of these meetings (scientific, NGOs, private sector, etc.). Since the Plenary meeting in Bonn (January 2013), a more formal process of developing the "stakeholder engagement strategy" was initiated and delegated to two organizations, i.e. IUCN (Box 8. 1) and ICSU (Box 8.2).[1]

Overall, 170 organizations participated in these meetings. Once the organizations were identified, we constructed a database to establish profiles based on one attribute, i.e. the organization type. Numerous categories may be found in the literature to distinguish between organization types. Some authors take the United Nations Framework Convention on Climate Change (UNFCCC) constituency categories into account (Lisowski, 2005; Vormedal, 2008) while others try to construct new categories through comparisons with UN major group classifications (Cabré, 2011).

Based on these works that analyzed stakeholder diversity in global environmental governance, we constructed a classification tailored to our sample and to our research concerns (Table 8.1) with seven categories. The first

Table 8.1 Distribution of stakeholder organizations by type

	Number of organizations
University/research	88
Environmental NGO	33
Development NGO	17
Indigenous people	12
Business	8
Others	7
Platform	5
Total	170

category is "university/research" (88 organizations). Historically, scientific networks and organizations are the first to get noticed in global environmental politics. In 1992, Peter Haas spoke of "epistemic communities" to describe networks "of professionals with expertise and competence recognized in a particular area and which claims authoritatively their political knowledge in this field" (Haas 1992: 3). ENGOs represent the second category and 33 are involved in the IPBES stakeholder group. ENGOs have been very active in global environmental governance since the Rio Earth Summit (1992). Twelve organizations representing indigenous or local communities are also involved as stakeholders. They were first mobilized at the national level and then tried to assert their claims in the international arena. Mobilization in international arenas is a way to enforce their demand at the national scale through a "boomerang effect", when their demands are not heard at lower scales (Keck and Sikkink, 1998). The business category is composed of eight organizations. Economic lobbies come later in the international environmental political game. This is not because they play a smaller role, but that they initially prefer involvement at the national level, rather than risking participation in international summits (Orsini, 2015). We noticed 17 development mandate oriented NGOs. They recently became involved in global environmental governance. Finally, we pooled seven organizations as "others", including specific professional sectors (forests, agriculture, engineering, legal) or emerging actors representing local authorities or urban areas. To complete our categories, we chose to identify platforms specialized on biodiversity issues (five organizations), these involve different kinds of organizations (governmental organizations, NGOs, scientific organizations, etc.). Platforms may not be numerous, but play a very specific role in the process of mobilizing stakeholders in IPBES at the regional scale.[2] These categories highlight the dynamics of power and influence among this diversity of stakeholders involved.

The distinction between these categories becomes fuzzier because the defended goals and interests of these organizations can also be shared by others. For example, the claims of indigenous groups in the negotiations of the Nagoya Protocol to the Convention on Biological Diversity adopted in 2010 were primarily driven by development NGOs such as the Third World Network. Their status is also rather ambiguous: the World Business Council for Sustainable Development (WBCSD) is not considered as an economic lobby but as an NGO, since it is a non-profit organization. Similarly, the Public Research and Regulation Initiative is officially a scientific lobby, while it is strongly sponsored by the economic lobby in favor of genetically modified organisms (GMOs) (Orsini, 2014a).

Some heterogeneous capacity to influence IPBES negotiations

Building on lobbying studies, the influence of stakeholders within a single forum depends on three interlinked resources: power (organizational, material, and ideational), combined with access and centrality (Orsini, 2013; Orsini, 2014b). The material and organizational power of business actors, such as Syngenta or WBCSD, which are involved in IPBES as stakeholders, gives them a strong capacity to influence decision makers in IPBES and beyond. Some environmental or research organizations can also have significant resources to influence decision makers. All of these kinds of organizations have different material, organizational, financial and ideational power, as Table 8.2 illustrates.

Different criteria are used in Table 8.2 to show examples of the unequal capacities of the stakeholders involved. The date of creation refers to the experience of the stakeholders and their legitimacy. The number of employees and the budget concern the organizational and financial capacities. Finally,[7] the number of publications highlights the influence of stakeholders on the scientific options and their credibility. However, the power of influence of these stakeholders must be analyzed through a combination of these different criteria and some other criteria, such as social capital (coalitions, etc.). So stakeholders that have few financial resources can compensate via some organizational and discursive capacities. According to Orsini (2014b), for example, indigenous groups have greater capacities to coordinate their claims than firms, which are more involved in competitive dynamics. Similarly, environmental NGOs are known for their discursive strategies, which allow them to communicate with civil society. Note that stakeholders also defend different interests and promote different frameworks regarding biodiversity, which could be in opposition. For example, according to business, biodiversity is presented as both an opportunity and a risk while, according to some scientists, biodiversity has an intrinsic value. For some indigenous people, biodiversity can also refer to property rights. The stakeholder group has become a new breeding ground to express these opposite interests and to balance power between business, scientific, indigenous and environmental organizations, etc.

Second, these IPBES stakeholders have not appeared out of nowhere. Most of them are also participating in other environmental regimes (CBD,

Table 8.2 Resources of some stakeholders

	Category	Date of creation	Number of employees	Budget	Number of publications in 2013
WWF	environment	1961	5,200	654 million[3]	141
DIVERSITAS	scientific	1991	5	1.04 million[4]	more than 500 publications of DIVERSITAS researchers
WBCSD	business	1996	64	3.9 million (euros)[5]	64
Tebrebba	indigenous	1996	37	4.2 million (euros)[6]	11
ICLEI (local governments for sustainability)	others	1990	220	12.5 million (euros)	5
Belgium biodiversity platform	platform	2001	12	700,000 (euros)	Not available
Rural Outreach Program Kenya	Development NGO	1992	2	79,667 (euros)	0

Source: adapted from Orsini, 2014b

IPCC, the Ramsar Convention, etc.) within the fragmented institutional context of global environmental governance (Biermann et al., 2009). We hypothesize that this participation in several forums is an advantage for IPBES stakeholders. More precisely, among these 170 organizations, we classified some which are also involved in other scientific arenas (IPCC and MA), and in some policy arenas (UNFCCC and the Conference of the Parties [COP] of the Convention on Biological Diversity). Among the 170 organizations, some have the capacity to circulate and evolve within IPBES and beyond as well as enhance the circulation of ideas, standards and knowledge. This situation can be an important advantage for some of them. They already have a professional network they can mobilize within IPBES, as well as links with delegations, etc.

Table 8.3 analyzes the involvement of the 170 organizations in four other environmental arenas (MEA, IPCC, COP of CBD, COP of UNFCCC). We identify one group of actors involved in all four forums (MA, IPCC, CBD and UNFCCC). Consequently, this group of actors (17 organizations) exhibits strong multi-positionality; they are referred to as the "Top 4" in the following analysis. At the other end of the spectrum, there is also a much larger group that is not connected to any of the four forums; they are referred to as "Isolated" (53 organizations).

A majority of organizations are involved in policy arenas (64%) with a smaller share involved in science arenas (35%). This highlights the fact that this majority is not particularly socialized in science arenas and its integration in the IPBES process could prove to be more difficult than expected. Among the organizations participating in science arenas, a large majority are focused on the climate (85%), and surprisingly, a smaller proportion on biodiversity (61%). While the presence of scientific organizations known for their involvement in biodiversity issues would guarantee continuity with the first scientific exercise in this field (MA), the predominance of organizations engaged in climate issues indicates a transfer of experience from IPCC to IPBES.

Finally, some stakeholders benefit from other channels to express themselves within IPBES or outside it. On the one hand, some stakeholders are close to ICSU or directly to Multidisciplinary Expert Panel (MEP) members and can express their claims through these channels. For example, some business organizations have favored links with some national delegations. Similarly, some scientific organizations and environmental organizations are very close to MEP. This is an important aspect insofar as some stakeholders "only" have the stakeholder days to express their viewpoints, whereas others benefit from one or more other channels (MEP, delegations, etc.) to make their position known within IPBES. On the other hand, some stakeholders can also protest against IPBES rules and express their claims outside of IPBES. Some organizations, e.g. the Friends of the Earth or the Society for Conservation Biology (SCB), are protest organizations that may be critical about the IPBES rules, and may even protest against them,[8] while at the same time taking part in the process.

Table 8.3 IPBES stakeholder participation in policy and science forums

	MA	IPCC	MA+IPCC*	MA+CBD	CBD	UNFCC	UNFCC +IPCC	CBD+UNFCC**	Top 4	Isolated	Total
University/ research	2	5	10	2	11	2	4	17	11	24	88
Environmental NGOs					6	5	1	7	5	9	33
Development NGOs						5				12	17
Indigenous people organizations				1	3			4		4	12
Business	1			1	2			2	1	1	8
Platform					2	1		1		1	5
Others						3		2		2	7
Total	3	5	10	4	24	16	5	33	17	53	170

** This sub-category of organizations concentrated in policy forums (CBD and UNFCCC) includes several organizations also linked to science forums (MA or IPCC).

IPBES stakeholder meetings: management and control of stakeholder interactions

This section analyzes the characteristics of the meetings that took place during the stakeholder days in Antalya (Turkey), in December 2013 (Pesche et al., 2014). The ethnographic analysis highlights the way the IPBES Secretariat organized, managed and controlled these meetings. This revealed both the supervisory process and the will to keep control over the stakeholder group, while also revealing the framework of interactions between the stakeholders present. Indeed, these interactions were fashioned, or even constrained, by the actual organization of the stakeholder days.

Formal management

The stakeholder days were held in a large conference room separate from the Plenary session room. At the Antalya session (IPBES-2), the stakeholder days benefited from considerable visibility (information panels, catering signage, etc.). Likewise, the layout of the room in which the event was held, its equipment in terms of lighting and image technology and logistical arrangement (presence of objects, room size, organization of seating, visibility of symbols, etc.) marked the desire to make it an important moment and thereby strengthen the interest of participants in the initiative. This arrangement also made it possible to control the nature of the stakeholder debates as the main event took place in the conference room, which thus presupposed the adoption of a tone that was polite, discursive and even technical. This formality and the seriousness of the moment were supported by the use of PowerPoint software and advanced lighting technologies. The layout of the conference room also gave a central role to facilitators of the stakeholder days, ICSU and IUCN representatives, with the help of two persons from the European Biodiversity Platform.

Controlled proceedings

The whole meeting lasted three half days, divided into three periods with: (1) a first phase devoted to an overview of IPBES, highlighting the still inclusive nature of the process ("new" people were welcomed and IPBES functioning explained to them); (2) a second phase devoted to the stakeholder involvement strategy, which was the subject of a specific debate during the Plenary session; and (3) a third phase aimed at preparing a dialogue with a few members of the MEP and the Bureau, and a joint stakeholder declaration.

After the introduction mentioned above, the first half-day was devoted to the general presentation of IPBES and the contents of the agenda for the second plenary session. Several speeches by representatives of the IPBES provisional secretariat (UNEP) provided a summary of IPBES functioning and of the expected role of stakeholders: a review of the background of IPBES and its links with other conventions and science-policy interfaces (IPCC and Millennium Ecosystem Assessment, etc.) helped politically and scientifically legitimize IPBES. Through its presentation format, this preliminary framework clearly illustrated the highly "productive" stakeholder participation expectations, which was mainly considered as a contribution to the Platform's work program.

A second session was organized with the aim of enhancing the IPBES stakeholder involvement strategy. The session was introduced by several people who took the floor, directed by the Platform. This notably included a female representative of indigenous peoples and a female representative from the private sector, and a representative from an association linked to the research world. It was then proposed to work in small groups on several topics in order to get participants acquainted with each other and facilitate discussion. A guideline document was given beforehand to the small groups of participants in this working session: "Considerations for the further development of a Stakeholder Engagement Plan (December 7–8, 2014)". This nine-page document included a table with a set of already formulated ideas on how stakeholders could contribute. Here again, the care taken to facilitate the meeting clearly showed the very close channelling of answers expected from participants at the stakeholder days. A few speakers questioned this process during the general debate: How was the panel of 1,500 people compiled to call upon the stakeholders? The organizers explained that, in 2010, they had assembled non-governmental stakeholder lists drawn up by IUCN and UNEP, in connection with the UN "major groups". Another question raised possible disagreements between stakeholders and a participant asked about the existence of a stakeholder database, and about the difference between observers, member countries and stakeholders. Without giving rise to any stormy debates, these few questions highlighted the, as yet, uncertain nature of the process and the mistrust mainly expressed by a small minority of participants.

The Third session, i.e. the final half-day of the stakeholder days, was devoted to preparing the dialogue with some official IPBES representatives from the MEP and the Bureau. A few of the MEP and Bureau members interacted with the floor concerning questions of governance and the role of stakeholders in the IPBES process. During the exchanges, indigenous group representatives marked their difference by calling for specific recognition as a partner group, as obtained under the Convention on Biological Diversity and in UN bodies in general (Coombe, 2001; Mauro and Hardison, 2000; Wallbott, 2014).

Most of the stakeholders agreed on acknowledging the importance of prior consultation and the fact that various consultations need to enable stakeholder viewpoints to be heard by governmental delegations. The issue of the diversity of viewpoints was touched upon again, countering the stated quest of the organizers for a common stakeholder position.

These features of the IPBES stakeholder days (place and proceedings) provided the framework for interactions which we shall now attempt to more effectively pinpoint.

Discursive style and dynamics of voicing viewpoints: avoiding disagreements and producing common sense

Unlike the IPBES Plenary sessions, there is no translation service offered during the stakeholder days and the working language is English, conveying a shared communication norm. Our ethnographic work also enabled us to analyze discursive styles, which were more or less formal or informal, ranging from the most antagonistic, where conflict seemed to be promoted, to the most conciliatory, where all trade-offs were possible to avoid conflict, to the most rational, where the best argument was generally sought. The observations of the Antalya stakeholder days highlighted the will to avoid disagreements and achieve a consensus among participants. The terms used were derived from management semantics: capacity building, benchmarking and consensus building. These terms have a "low emotive value, but a strong instrumental and functional value by avoiding any ideological or political reference" (Kalaora, 1999). The rhetoric used by the organizers was not dissuasive, even when the indigenous representatives wished to be considered as partners. The rhetoric used did not seek to impose a single vision either, but sought above all to find a common meaning on which all could agree. The stakeholder day meetings served somewhat as a secondary socialization event (Berger and Luckmann, 1991) making it possible to adopt new standards and rules in line with the expectations of the group. An international meeting like that of the IPBES stakeholder days requires demonstrations of respect and courtesy at all times, especially because nobody knows where the acceptable limits of verbal argumentation lie for their partners. Through their participation, stakeholders who had not yet been socialized to these standards learnt the rules of conduct (Van Vree, 2001). Some "precise, constant and smooth behaviors" are instilled, thus constraining expressions of affect and emotion.

Any negotiation on content required negotiation on the procedure, on what Foucault called "the management of management" (Foucault, 2014). This sometimes slows the process down considerably, which may be exasperating for observers. The issue of the slowness of international negotiations is well documented in the scientific literature (Abélès, 2008; Bendix, 2012; Kelly, 2009; Müller, 2012; Riles, 1999) and notably highlights that in the same place – in this case the conference room of the stakeholder days – heterogeneous types of participants, of political and geographical contexts, and of knowledge are brought together. Assembled for two days, they represented places all over the world. Participants thus took home, i.e. disseminated worldwide, what was discussed and re-discussed in the form of successive draft wordings during the meetings. Moreover, the heterogeneity of participants, who did not share a common interest, also further slowed down the work of the stakeholders.

Conclusion

The stakeholder group has become a new space to express a range of interests and balance the power between different organizations. It is composed of heterogeneous members, including scientific associations, business organizations, indigenous communities and environmental organizations which have unequal capacities to influence the IPBES process. This depends on organizational, material and ideational resources, as well as centrality. We noted that few IPBES stakeholders had strong multi-positionality in global climate and biodiversity forums, while another group of 53 members were considered as being isolated. Due to this heterogeneity, the functioning of the stakeholder group imposes a way of participating and a precise discursive style, which the participants are encouraged to follow. It is therefore within this framework proposed by the organizers of the stakeholder days that members can express themselves.

This leadership is designed for utilitarian ends: favoring expressed viewpoints that converge with the organizers' expectations, or seeking to exclude or minimize expressed positions that would not seem to comply with them.

The involvement of some organizations was very close to meeting the organizers' expectations. For example, specific environmental organizations fully accepted the rules of the game and the organizers' objectives and fully took part, e.g. the World Wide Fund for Nature (WWF) members. On the contrary, this leadership, in seeking to exclude or minimize expressed positions that would be out of line with the mainstream agenda could lead to the exclusion of some organizations. For instance, Indigenous and Local Communities (ILCs) were considered important in the IPBES Plenaries because of their non-scientific knowledge inclusion potential (Buizer, Arts and Kok, 2011; Opgenoorth and Faith, 2013; Thaman et al., 2013; Turnhout et al., 2012). But ILC representatives seldom voiced their views over the two days of the meetings – instead they wanted to be considered as partners, not as stakeholders like the others. That claim reflected the status of indigenous peoples in CBD who have managed, since the mid-1990s, to be acknowledged as full-fledged actors in negotiations. Their stakeholder status in IPBES was seen as a step backwards and a way of minimizing their ability to produce knowledge considered as legitimate.

Furthermore, private sector representatives as well as some scientific experts, for example, had other channels for expressing their position within IPBES and therefore they accepted to follow the rules of the game of the stakeholder group. They know that they have to be there to be legitimate, but they will try to influence decision making through other informal channels (by influencing government delegates or MEP members) or/and formal channels (suggesting experts for IPBES assessments). It is not the same situation when an organization and its members know that the stakeholder group is the only channel they have. Then they need to change the rules which could be too strict and controlled, or otherwise leave the whole process.

The idea of a group of stakeholders, as promoted by IPBES, is accompanied by a leadership/participation model that tends to encourage stakeholders to get involved or, conversely, lead them to withdraw from the stakeholder dynamics. The most contentious and heterogeneous categories might then prefer to rule themselves out.

Notes

1 Before 2013, UNEP established agreements with ICSU and IUCN. The aim of the agreement with the first institution was to foster the involvement of the scientific community in IPBES, while the second was dedicated to pushing for the engagement of NGOs and trying to reach the private sector and indigenous and local communities.
2 We removed the few international organizations and conventions that were involved in the stakeholder days because of their special position as "strategic partners" in IPBES governance (see chapter 4 on IPBES governance).
3 World Wildlife Fund Annual Report 2013.
4 DIVERSITAS Annual Report 2013.
5 World Business Council for Sustainable Development Annual Report 2013.
6 Tebtebba Annual Report 2013 (Indigenous Peoples' International Centre for Policy Research and Education).
7 Other criteria such as the number of members (organizations or individuals) and the number of linked organizations can also be used to analyze the unequal capacities (see Orsini, 2014a).
8 For instance, some SCB representatives were very vocal about the importance for stakeholders to nominate members of the IPBES Multidisciplinary Expert Panel.

References

Abélès, M. 2008. *Anthropologie de la globalisation: débats et perspectives*. Paris: Payot.

Alcock, F. 2008. Conflicts and Coalitions within and across the ENGO Community. *Global Environmental Politics*, 8, 66–91.

Arts, B. 2005. Non-state Actors in Global Environmental Governance: New Arrangements beyond the State. *In*: Koenig-Archibugi, M. & Zürn, M. (eds.) *New Modes of Governance in the Global System*, Basingstoke: Palgrave Macmillan.

Bäckstrand, K. 2006. Democratizing Global Environmental Governance? Stakeholder Democracy after the World Summit on Sustainable Development. *European Journal of International Relations*, 12, 467–498.

Bendix, R. 2012. Une salle, plusieurs sites: les négociations internationales comme terrain de recherche anthropologique. *Critique internationale*, 19–38.

Berger, P. L. & Luckmann, T. 1991. *The Social Construction of Reality: A Treatise in the Sociology of Knowledge*, London: Penguin UK.

Bernstein, J. 2010. Assessing the Value of Civil Society Involvement in IPBES Governance. Briefing Paper. IUCN.

Bernstein, S. & Cashore, B. 2007. Can Non-state Global Governance be Legitimate? An Analytical Framework. *Regulation & Governance*, 1, 347–371.

Betsill, M. M. & Corell, E. 2008. NGO Diplomacy: The Influence of Nongovernmental Organizations in International Environmental Negotiations, Cambridge, MA: MIT Press.

Biermann, F., Pattberg, P., Van Hasselt, H. & Zelli, F. 2009. The Fragmentation of Global Governance Architectures: A Framework for Analysis. Global Environmental Politics, 9, 14–40.

Buizer, M., Arts, B. & Kok, K. 2011. Governance, Scale and the Environment: The Importance of Recognizing Knowledge Claims in Transdisciplinary Arenas. Ecology and Society, 16.

Cabré, M. M. 2011. Issue-linkages to Climate Change Measured through NGO Participation in the UNFCCC. Global Environmental Politics, 11, 10–22.

Clapp, J. 2005. Global Environmental Governance for Corporate Responsibility and Accountability. Global Environmental Politics, 5, 23–34.

Conca, K. 1995. Greening the United Nations: Environmental Organisations and the UN System. Third World Quarterly, 16, 441–458.

Coombe, R. J. 2001. Recognition of Indigenous Peoples' and Community Traditional Knowledge in International Law. Thomas Law Review, 14, 275.

Corell, E. & Betsill, M. M. 2001. A Comparative Look at NGO Influence in International Environmental Negotiations: Desertification and Climate Change. Global Environmental Politics, 1, 86–107.

ENB 2008. IPBES: A Report of the Ad hoc Intergovernmental and Multi-Stakeholder Meeting on an Intergovernmental Science-Policy Platform on Biodiversity and Ecosystem Services (IPBES). Earth Negotiations Bulletin, International Institute for Sustainable Development (IISD).

ENB 2015. IPBES Bulletin. Summary of the Stakeholder Days Prior to the Third Session of the Intergovernmental Platform on Biodiversity and Ecosystem Services (IPBES-3). Earth Negotiations Bulletin, International Institute for Sustainable Development (IISD).

Foucault, M. 2014. Dits et écrits (1980–1988), Paris: Gallimard.

Freeman, R. E., Harrison, J. S., Wicks, A. C., Parmar, B. L. & De Colle, S. C. (eds.) 2010. Stakeholder Theory: the State of the Art, Cambridge: Cambridge University Press.

Haas, P. M. 1992. Introduction: Epistemic Communities and International Policy Coordination. International Organization, 46, 1–35.

Haufler, V. 2009. Transnational Actors and Global Environmental Governance. Governance for the Environment: New Perspectives, 119–143.

Interacademy Council 2010. Climate Change Assessments. Review of the Processes and Procedures of the IPCC. IAC, Committee to Review the Intergovernmental Panel on Climate Change.

IPBES 2014. Revised Draft Stakeholder Engagement Strategy (deliverable 4 (d)) (IPBES/3/16). United Nations Environment Programme (UNEP).

Kalaora, B. 1999. Global Expert: La religion des mots. Ethnologie française, 513–527.

Keck, M. E. & Sikkink, K. 1998. Activists beyond Borders. Advocacy Networks in International Politics, Ithaca: Cornell University Press.

Kelly, T. 2009. The UN Committee against Torture: Human Rights Monitoring and the Legal Recognition of Cruelty. Human Rights Quarterly, 31, 777–800.

Kowalski, A. A. & Jenkins, L. D. 2015. The Role of Bridging Organizations in Environmental Management: Examining Social Networks in Working Groups. Ecology and Society, 20, 16.

Lisowski, M. 2005. How NGOs Use Their Facilitative Negotiating Power and Bargaining Assets To Affect International Environmental Negotiations. *Diplomacy and Statecraft*, 16, 361–383.

Mauro, F. & Hardison, P. D. 2000. Traditional Knowledge of Indigenous and Local Communities: International Debate and Policy Initiatives. *Ecological Applications*, 10, 1263–1269.

Mitchell, R. B. 2010. *International Politics and the Environment*, New York: Sage Publications.

Müller, B. 2012. Comment rendre le monde gouvernable sans le gouverner: les organisations internationales analysées par les anthropologues. *Critique internationale*, 54, 9–18.

Opgenoorth, L. & Faith, D. P. 2013. The Intergovernmental Science-policy Platform on Biodiversity and Ecosystem Services (IPBES), Up and Walking. *Frontiers of Biogeography*, 5.

Orsini, A. 2013. Multi-Forum Non-State Actors: Navigating the Regime Complexes for Forestry and Genetic Resources. *Global Environmental Politics*, 13, 34–55.

Orsini, A. 2014a. Les lobbies environnementaux: intérêts d'une approche pluraliste. *CERISCOPE Environnement*. http://ceriscope.sciences-po.fr/environnement/content/part3/les-lobbies-environnementaux-interets-d-une-approche-pluraliste (last accessed 1 June 2016).

Orsini, A. 2014b. The Role of Non-state Actors in the Nagoya Protocol Negotiations. In: Oberthür, S., Rosendal, G. K. (eds.) *Global Governance of Genetic Resources: Access and Benefit Sharing after the Nagoya Protocol*. Abingdon: Routledge, 60–78.

Partridge, K., Jackson, C., Wheeler, D. & Zohar, A. 2004. *The Stakeholder Engagement Manual. Volume 1: The Guide to Practitioners' Perspectives on Stakeholder Engagement*. Stakeholder Research Associates Canada Inc., UNEP.

Pattberg, P. & Stripple, J. 2008. Beyond the Public and Private Divide: Remapping Transnational Climate Governance in the 21st century. International Environmental Agreements. *Politics, Law and Economics*, 8, 367–388.

Pesche, D., Oubenal, M., Vandevelde, J.-C. & Hrabanski, M. 2014. Le consensus d'Antalya: les avancées de la plateforme science-société sur la biodiversité et les services écosystémiques (IPBES). *Nature, sciences et sociétés*, 22, 240–246.

Riles, A. 1999. Models and Documents: Artefacts of International Legal Knowledge. *International and Comparative Law Quarterly*, 48, 805, 808.

Thaman, R., Lyver, P., Mpande, R., Perez, E., Cariño, J. & Takeuchi, K. 2013. The Contribution of Indigenous and Local Knowledge Systems to IPBES: Building Synergies with Science. IPBES Expert Meeting Report. Paris: UNESCO/UNU.

Turnhout, E., Bloomfield, B., Hulme, M., Vogel, J. & Wynne, B. 2012. Conservation Policy: Listen to the Voices of Experience. *Nature*, 488, 454–455.

UNEP 2009. Summary of Perspectives from the Scientific Community and Broader Civil Society (UNEP/IPBES/2/INF/4). Nairobi.

UNEP 2010. Options for Improving the Science-policy Interface for Biodiversity and Ecosystem Services (UNEP/IPBES/3/2).

UNEP 2013. Draft Stakeholder Engagement Strategy for Supporting the Implementation of the Platform's Work Programme (IPBES/2/13). United Nations Environment Programme (UNEP).

UnitedNations 2004. We the Peoples: Civil Society, the United Nations and Global Governance. Report of the Panel of Eminent Persons on United Nations–Civil Society Relations. New York: United Nations.

Van Vree, W. 2001. *Meetings, Manners, and Civilization: The Development of Modern Meeting Behaviour.* Leicester University Press.

Vormedal, I. 2008. The Influence of Business and Industry NGOs in the Negotiation of the Kyoto Mechanisms: The Case of Carbon Capture and Storage in the CDM. *Global Environmental Politics*, 8, 36–65.

Wallbott, L. 2014. Indigenous Peoples in UN REDD+ Negotiations: "Importing Power" and Lobbying for Rights through Discursive Interplay Management. *Ecology and Society*, 19, 21.

9 Capacity building for global science-policy interface activities

The establishment of the IPBES task force on capacity building

Sélim Louafi

Introduction

Capacity building in a global setting is often the only thing that everyone agrees is important and needed. Such a broad consensus is usually not a good sign. The term ranges from a very narrow vision, often associated with training activities and workshops, to an all-encompassing vision that tends to be useless from an analytical and practical standpoint (Schacter, 2000). It could mean helping respond to a lack of technical or scientific skills, of money, time or authority to do all the things expected, or to a lack of institutional capacity (Potter & Brough, 2004). It has been and continues to be a major component, if not a motto, of many global development programs, as well as a major component of any international documents and protocols related to global change and sustainable development (UNEP, 2005). Its importance has also been perceived as critical in science-policy interface discussions at the global level (Kleine, 2009). The need to increasingly address challenges of global scale calls for science to play a central role in the making of global policy (Miller, 2001). This increased importance of science in formulating global policy, however, raises several concerns, particularly with regard to the unequal capacity of countries to contribute to the production of this science. Capacity building plays an even more important role in this context. Developing countries repeatedly insist on the need to increase their active participation in scientific research and monitoring, global scientific assessments, and on the need to build enough capacity for the formulation of national science policies and related action plans.

But the issue goes even beyond science to encompass all modes of knowing and deciding. The basic structure of the biodiversity problem raises challenging questions with regard to the types of knowledge that "count" and/or are necessary for decision making. Dealing with biodiversity issues entails a multiplicity of legitimate perspectives and discourses laden with conflicts over facts, interests and values (Koetz, Farrell & Bridgewater, 2012). As highlighted by the experience of the Millennium Ecosystem Assessment (Miller & Erickson,

2006), building a credible, legitimate and salient assessment on biodiversity requires substantial knowledge and expertise pooled by various actors at various scales. This includes scientific knowledge, but also traditional knowledge and practitioners' knowledge that "often dominates the considerations of site-specific resource management issues, where detailed scientific studies may not exist" (Reid et al., 2006: 2).

Tensions and conflicts also emerge over the appropriate level of institutional intervention for biodiversity conservation and sustainable use. The global dimension of the biodiversity problem is often questioned as biodiversity loss and changes in ecosystem services are typically place-based and many of the effects are seen at sub-global scales (Swanson, 1999; Duraiappah & Rogers, 2011). Yet, the global biodiversity issue is more than the sum of all local biodiversity crises as it also involves a systemic dimension linked to global population growth, production and consumption patterns or global land-use patterns. Biodiversity issues also encompass various levels of observation of living systems – from genes to ecosystems through species – and associated traditional knowledge and know-how associated with each of these levels. The capacity to account for the complex set of interlinkages within and across these levels is at the core of the biodiversity concept. Yet discussions are still very often fragmented across the various disciplines associated with these various levels.

Biodiversity spans sectors of human activities and relates to a broad set of values, from utilitarian use values related to the direct or indirect benefits that humans gain from ecosystem goods and services to non-utilitarian ethical or stewardship values (Van den Hove & Chabason, 2009). At least five interconnected elemental issue-areas account for the different facets through which global biodiversity governance is approached by the various actors and institutions (Swanson, 1999; Swanson & Groom, 2012; Rosendal, 2001; Le Prestre, 2002; Brahy & Louafi, 2004; Brand & Gorg 2013; Raustiala & Victor, 2004; Morin & Orsini, 2014): conservation, development, trade, culture and agriculture. These issues have emerged at the global level at different periods, pushed by different community actors and embodied in different institutions.

What role does capacity building have in this setting of political tensions with regard to the way biodiversity is approached at different scales by various stakeholders, academic disciplines and sectors? This chapter describes the way this question has been framed and addressed within the International Platform on Biodiversity and Ecosystem Services (IPBES). Starting with the current IPBES mandate and activities in terms of capacity building, the chapter develops a conceptual framework on capacity building that is used to boost our insight on the dynamics (content-wise) that have led to such an outcome in terms of capacity building. We ask: What were the capacity building objectives at the outset of the IPBES establishment process? How have these objectives evolved? What was left aside? What was considered as critical and why?

Capacity building activities within IPBES

In establishing IPBES, governments recognized that

> the Platform *prioritizes* key capacity building needs to *improve the science-policy interface at appropriate levels* and then *provides* and *calls for financial and other support* for the highest priority needs related to its activities, as decided by the Plenary, and *catalyses financing for such capacity-building activities* by providing a forum with conventional and potential sources of funding.[1]
>
> (UNEP, 2012: 10)

This key statement already highlights important decisions that have been made with regard to capacity building. First, three types of activities are identified: prioritizing capacity building needs; providing support; and catalyzing new funds. Second, capacity building is targeted to science-policy interface activities by opposition, for example, to capacity building for biodiversity and ecosystem services in general. Although the scope is potentially enormous, this narrows down the type of activities that should be covered. Third, the scale challenge is highlighted ("at the appropriate scale"). As discussed in the introduction, it reflects the specific challenge of biodiversity loss, whose effects are felt at sub-global levels and, subsequently, the tensions around the appropriate level of institutional intervention for biodiversity conservation and sustainable use. Fourth, it recognizes that priority in terms of support should be allocated to the activities of the Platform, as decided by the Plenary, while other activities should benefit from new mechanisms to attract and leverage new funding.

These objectives have been translated into different operating principles which are intended to guide the Platform regarding the manner it carries out its work (UNEP, 2010a; IPBES, 2014). These operating principles reflect various issues that were critical in the decision to establish IPBES. First, it is recognized that when it comes to capacity building, IPBES does not operate in a vacuum and should "collaborate with existing initiatives,[2] to fill gaps and build upon their work while avoiding duplication" (IPBES, 2014: 2). Second, these operating principles also reflect the difficult equilibrium that needs to be achieved in any science-policy interface – the process of information and knowledge generation should be built on classical science norms, while also relying on the political (democratic) principles of inclusiveness and transparency: to "use clear, transparent and scientifically credible processes for exchange, sharing and use of data, information and technologies from all relevant sources" and "recognize the need for full and effective participation of developing countries, and balanced regional representation and participation" and "recognize and respect the contribution of indigenous and local knowledge to the conservation and sustainable use of biodiversity and ecosystems" (Ibid.). Third, these principles speak also to the scientific community by recognizing the need to: "take an interdisciplinary and multidisciplinary approach that incorporates all relevant disciplines including both social and natural sciences". Finally, they also recognize the specific challenge of avoiding the isolation of capacity building activities and integrating them "into all relevant aspects of the Platform work according to priorities identified by the Plenary" (Ibid.).

A task force devoted to capacity building (CB) was established following the second IPBES Plenary in December 2013. It consists of two Bureau members and three members of the Multidisciplinary Expert Panel (between them covering the five United Nations regions) and 20 capacity building experts from various regions worldwide selected by the IPBES Bureau and the Multidisciplinary Expert Panel. These experts are mainly from government agencies (mainly ministries of environment), national or regional biodiversity agencies (e.g. Wildlife Institute of India, South African National Biodiversity Institute, ASEAN Centre for Biodiversity, etc.), or university and research institutes. It is headed by the IPBES Chair Prof. Zakri Abdul Hamid of Malaysia, and IPBES Bureau member Mr. Ivar Baste of Norway. It aims to support delivery of the elements of the work program related to CB. A technical support unit provided by the Government of Norway, provides technical assistance to this task force.

Based on the terms of reference[3] agreed by the Second Plenary in its decision IPBES-2/5, the task force has developed its activities in the four following areas: 1) Identifying and prioritizing capacity building needs; 2) Partnership, exchange and training programs; 3) Increasing access to technical and financial resources; 4) Building and enabling networks to address capacity building needs. The work mainly takes place through face-to-face meetings (the task force meets about twice a year) based on input documents prepared by the Secretariat and the Technical Support Unit. During the meetings, work is usually divided by break-out groups in order to come up with concrete outputs more efficiently. Inter-sessional work is carried out mainly through electronic interactions where experts are asked to provide feedback on follow-up actions implemented by the Technical Support Unit or, possibly, by other members or experts invited by the task force.

A framework for analyzing capacity building

In the academic literature, capacity building is not clearly associated with a specific discipline. So many elements could fall under capacity building depending on the discipline and the sector of interest and there is no universal theory for approaching capacity building. Some authors insist on the capacity building process, others on its components or the mechanisms and tools needed to achieve it (Franks, 1999; Kleine, 2009). For the purpose of this chapter, building on the work of Moore, Severn and Millar (2006) and Robins (2008), we distinguish four broad elements of capability that make up capacity (human capital, social capital, institutional capacity and economic capital) as an analytical device to facilitate structured discussion around capacity building. These four elements of capability cover the various scales at which capacity building could intervene: individual, community, institutional as well as national or regional (see Table 9.1).

First, capacity building activities devoted to human capital include initiatives that aim at increasing individual skill, experience and knowledge, but also wealth. It involves education, but also acquisition of skills and know-how through job or life experiences. Other broader capacities such as leadership and capacity to engage and innovate are also important parts of human capital. Second, social capital refers to both social norms (such as trust-building and reciprocity and more generally to values, attitudes and behaviour, commitment and motivation) and structural aspects of relationship and collaboration building, such as networks, collective action and association and information flows. Third, institutional capital refers to the formal rules that allow individuals, organizations, and society as a whole to interact and properly function (Moore, Severn and Millar, 2006; Robins, 2008). This includes, *inter alia*, legal, regulatory and administrative rules, human resource development, management practices and procedures, including monitoring and control systems. Fourth, economic capital includes physical (infrastructure) and liquid (financial) resources.

These four broad categories are, however, too broad to help identify more specific challenges related to the science-policy interface. Based on the objectives developed by Robins in her capacity building framework and on a literature review regarding the science-policy interface, we propose a set of six clusters that better specify potential measures and scales of intervention (individual, community, institutional, national). These clusters are: 1) Grow the knowledge base; 2) Build the information base and tools; 3) Develop and sustain networks and relationships; 4) Strengthen institutional science-policy interface structures and processes; 5) Access infrastructure and resources; 6) Define/refine the organizational business (see Table 9.2).

Table 9.1 A conceptual framework for facilitating structured discussion about capacity development

Human capital	Social capital		Institutional capital	Economic capital
	Social norms	Structural		
Knowledge	Trust and reciprocity	Networks	Governance arrangements	Infrastructure
Skills	Values, attitudes, and behavior	Relationships		Financial resources
Experience	Commitment Motivation			

Source: Modified from Robins (2008)

Table 9.2 Clusters of capacity building objectives for the science-policy interface

	Human capital	Social capital	Institutional capital	Economic capital
A. *Grow the knowledge base*	++	+		
This cluster mainly includes human capital enhancement measures such as competency-based training, personal and professional development and mentoring and coaching				
B. *Build the information base and tools*				
This cluster also focuses on human capital enhancement by increasing the generation, collection, collation, organization, exchange, use and analysis of information as well as decision-making support systems	++	+	+	
C. *Develop and sustain networks and relationships*				
This cluster mainly deals with social capital enhancement through mechanisms for establishing and developing relationships, networks, teams, and long term partnerships among stakeholders and measures such as recognition and awards. It also includes mechanisms for improving communication products tailored to the need of various stakeholders	+	++		
D. *Strengthen institutional science-policy interface structures and processes*				
This cluster mainly covers social and institutional capital enhancement through legal tools (protocols), policy mechanisms (multi-stakeholder platform, national coordination bodies, brokering organizations, communities of practices, implementation platform) and regulatory tools (assessment, monitoring)		++	++	
E. *Access infrastructure and resources*				
This cluster mainly deals with improvement of the infrastructure and financial dimensions of institutional capacity. It includes measures such as training facility, information systems and flexible and adequate financial arrangements (matchmaking facility)			++	++
F. *Define/refine the organizational business*				
This cluster mainly covers any reflexive measures on capacity building aimed at better fulfilling the organization's missions. It includes measures such as identification of priority needs and existing capacities, best practices, performance measures and reporting, evaluation and feedback processes.			++	

These clusters all contribute directly or indirectly to enhance the science-policy interface. They should however not be seen as part of a linear process but rather as an intertwined set of objectives that reinforce each other (see Figure 9.1). The measures within each cluster generally contribute to one previously identified capital category but, depending on the scale of intervention, could also contribute to more categories.

A background analysis on the establishment of the capacity building task force

This background analysis is based on an in-depth review of the various documents produced as part of the negotiation process for the establishment of IPBES since its inception in 2005 with the consultations held under the International Mechanism of Scientific Expertise on Biodiversity (IMoSEB). The IISD reports[4] for these meetings have been used, as well as direct interviews with a few key players in this negotiation process. The framework developed in the previous section is used as the basis for conducting the analysis in order to better understand the development of the capacity building discussion and its outcome, as reflected in the task force on capacity building mandate.

The analysis distinguishes three phases: the first phase, from 2005 to 2007, was mainly devoted to consulting various stakeholders located in various regions worldwide to reach a consensus on the need to establish a science-policy

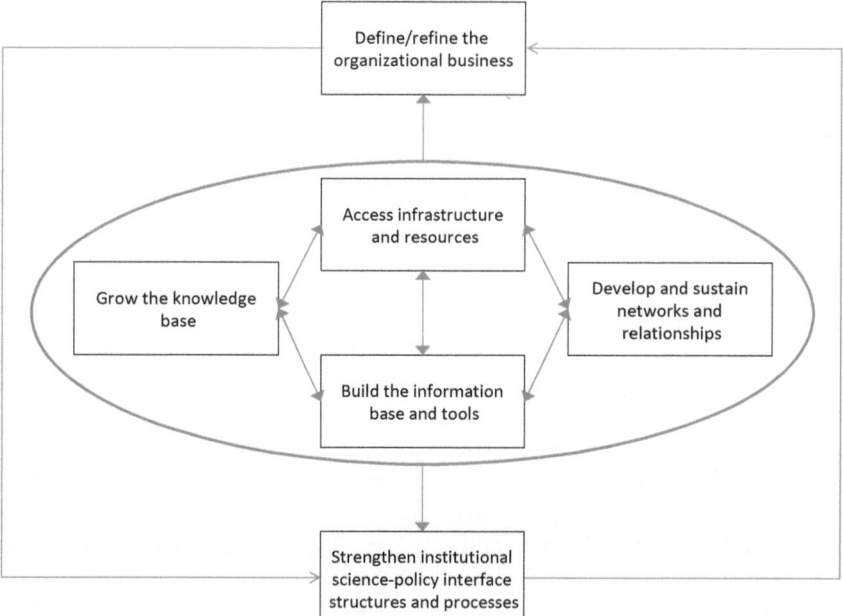

Figure 9.1 Clusters of capacity building objectives for the science-policy interface

interface at the global level. Capacity building issues were identified as an important dimension and covered in quite general terms; the second phase, from 2008 to 2010, involved negotiation on initiatives such as the IPBES establishment. Capacity building principles were agreed and specific discussions on potential measures initiated. The third phase, from 2011 to 2013, was devoted to definition of the Platform mandate and functions, establishment of the multidisciplinary expert panel and its task forces.

The first phase (2005–2007): the need for a science-policy interface

This first phase mainly involved a wide consultation with relevant stakeholders organized at regional levels to discuss the need for establishing a science-policy interface body at the global level (Babin et al., 2008). The discussion focused mainly on the potential mandate of such a body and collected views on various issues, including capacity building. However, the discussion remained very general on this matter.

The IMoSEB consultation process started with the meeting in North America and subsequent meetings were conducted in Africa, Europe, Asia, and South America in 2007. During these meetings, participants discussed challenges related to biodiversity and ecosystem services. It was noted that, while developing countries often have a high level of biodiversity, they have to cope with many challenges, such as poverty, sanitation and population growth, which have negative effects on sustainable development and the environment. Thus, most countries acknowledged the need for capacity building in developing countries, particularly to develop the capacity to predict consequences of current actions on biodiversity, and effectively assess threats related to biodiversity. This conception of capacity building in terms of improvement of scientific knowledge and enhancement of the capacity to deliver scientific information to decision makers relies on a linear model of expertise, where science speaks truth to power (Pielke 2007; Beck, 2010) and where a science-policy interface fulfils the role of a translation body that enhances communication and consensus between the two distinct communities of science and politics (Vadrot, 2013). This first conception that was at the core of the launching of the initiative during the Science and Biodiversity Conference in Paris in 2005, has been progressively challenged during the IMoSEB consultation process by another vision that focuses more on interface activities with the aim of simultaneously building knowledge capacities for the providers and recipients. It considers biodiversity erosion as mainly the result of conflicting perceptions and views about the very issue of biodiversity and how it should be preserved (Miller & Erickson, 2006; Koetz, Farrell & Bridgewater, 2012). Acknowledging this plurality of knowledge forms and political perceptions calls for a flexible, multilayered, bottom-up science-policy interface design (Vadrot, 2013) and hence capacity building activities that focus on enhancing mutual understanding in order to increase the cooperative capacities of the various actors in addressing collectively agreed objectives.

This major divide has many concrete implications that will structure the discussions during this first phase but also the subsequent ones. Three issues are particularly interesting to note:

- The appropriate scale for capacity building activities. This issue was part of the broader discussion about the very need for a science-policy interface body on biodiversity and ecosystem services at the global level or for a more decentralized bottom-up approach. This question of the appropriate geographical scale of intervention also relates to the target level of capacity building interventions, from individual to national or regional.
- The level of connection to and embeddedness within existing global organizations or policy processes, in particular CBD. It was noted that the decision-maker target audience should be broader than the international conventions related to biodiversity (e.g. indigenous people and local communities, private sector, NGOs, academia, media) and that independence should be ensured for a well-functioning platform. This discussion builds upon some dissatisfaction with the CBD Subsidiary Body on Scientific, Technical and Technological Advice (SBSTTA) presented as "an open-ended intergovernmental scientific advisory body to provide the Conference of the Parties (COP) and, as appropriate, its other subsidiary bodies, with timely advice relating to the implementation of the Convention" (art. 25 of the Convention of Biological Diversity). Quite soon after its formal commencement, SBSTTA started to be criticized for its functioning as a mini-CBD committee dealing with science and technology matters rather than as a true science-policy interface body able to produce agreed scientific-based evidence and assessments about biodiversity erosion on which to base policy decisions (see Koetz et al., 2008).[5] At the other extreme, global assessment efforts, such as the Global Biodiversity Assessment or the Global Biodiversity Outlook, were also heavily criticized for their purely scientific-driven processes and for not paying enough attention to political and social dimensions of biodiversity issues (Duraiappah & Rogers, 2011). As noted by Koetz, Farrell & Bridgewater (2012), distributing authority, resources and capacities across multiple institutions, rather than restricting them to a single central global authority, could improve relevance, credibility and legitimacy in dealing with the complexities of global biodiversity governance challenges.
- The level of embeddedness of capacity building activities within the work program of the future science-policy interface body. Conflicting views were expressed on whether capacity building activities should be integrated within the work of the future Platform or as a separate set of activities. This issue directly relates to what is presented as the last cluster (defining/refining the organizational business) in our conceptual framework: Is the role of IPBES to push for developing capacities on

biodiversity issues as one among other kinds of activities it could undertake or to integrate it in all of its own activities? This latter option was considered critical by many developing countries to ensure their full participation in all activities that the future science-policy interface would undertake.

Second phase (2008–2010): agreement on capacity building principles and initial mapping of capacity building activities

The 2008 to 2010 meetings were *ad hoc* intergovernmental and multi-stakeholder meetings aimed at finding an agreement on the need and possible mandate for a new organization that would act as a science-policy interface for biodiversity.

Following one of the issues raised during the first phase, three potential approaches for dealing with capacity building were discussed during the 2009 meeting:

1 Externalizing capacity building activities by increasing capacity building with regard to cooperation, assessments and policy implementation of initiatives under various international organizations;
2 Establishing a new mechanism that supports existing capacity building initiatives; and
3 Completely internalizing capacity building activities within the work program of the new science-policy interface to be created.

The meeting did not take a final position, but a majority of voices favored the third option. Within the governance structure agenda item, the need to establish a permanent working group to specifically deal with capacity building activities was even recognized. Besides, the meeting recognized

> the importance of capacity building for the generation, assessment and use of knowledge at various levels. Capacity building for scientists, policy-makers and members of civil society, including local communities, should be catalyzed to enable them to participate more effectively in the science-policy interface, in addition to increasing the participation and involvement of scientists from developing countries and ensuring that focused technical and scientific support was provided to facilitate that greater involvement.
>
> (UNEP, 2009: 8)

The meeting also started collecting views on specific measures, but requested additional analysis to be provided with regard to current and planned capacity building activities.

Such analysis was presented as an information document[6] by the Secretariat at the third meeting held in 2010 in Busan (South Korea), with the following title: "Analysis of capacity development for biodiversity and ecosystem services". This document is crucial as it proposes, for the first time, a concrete and systematic way of approaching capacity building based on a conceptual framework and a description of potential measures to be included in the future science-policy interface body. The document also proposes a definition of capacity development and discusses the role of diverse stakeholders in capacity building and dissemination of information. The document builds upon the results of a questionnaire that queried countries about their capacity development activities; the analysis of obstacles to the implementation of national biodiversity strategy and action plans; and national capacity self-assessment results from the Capacity Development Initiative of the Global Environmental Fund and the United Nations Development Programme.

Interestingly, the document develops a broad perspective on capacity building as it does not only focus on science-policy interface activities but also encompasses all potential activities for biodiversity and ecosystem services. It also takes a strong stand from the outset, as reflected in the title and on the first page, by explicitly making reference to capacity development by opposition to capacity building. The document notes that

> [t]he terms 'capacity-building' and 'capacity development' are often used interchangeably. As used in the present document and by most international donors and the Organization for Economic Cooperation and Development, for example in the Paris Declaration on Aid Effectiveness, the two terms are in fact different. Capacity development denotes a relatively long-term process that aims at supporting governance structures in efforts to become self-reliant and capable of better delivering development results. Capacity-building, on the other hand, suggests a relatively short-term and more technical approach, particularly targeting individual capabilities through training.
>
> (UNEP, 2010b: 3)

This distinction between long term/structural support and short term/technical support targeted to individual capacities has not sunk in and, probably by institutional inertia rather than through a conscious decision, IPBES has kept the expression "capacity building" in its broad sense, i.e. without formalizing the proposed distinction. Note also that this binary distinction between what can be referred to as institutional vs human capital in our framework, does not really account for other important capacity dimensions such as social capital, yet it is essential for science-policy interface activities to increase mutual understanding among and within various groups. It also does not explicitly incorporate economic capital.

Second, the document proposes a framework for "analyzing achievements and gaps in the generation and use of scientific knowledge and assessments, which were identified as key areas for the science-policy interface". This framework considers capacity as a set of interactive functions within five functional capacity clusters required to manage and achieve specific objectives:

a   Capacity to engage stakeholders;
b   Capacity to gain access to and use information and knowledge;
c   Capacity to plan processes and develop policy;
d   Capacity to manage and implement;
e   Capacity to monitor and evaluate.

Although the importance of collaboration among stakeholders is emphasized, the framework remains largely focused on capacity building activities at the national level and the development of a linear science-policy interface approach. The specific ordering of these clusters suggests that specific knowledge, when properly accessed and used, leads automatically to better policy development and implementation (Pielke, 2007; Koetz, Farrell & Bridgewater, 2012).

However, when looking in more detail at the description of these clusters, it appears that this is more an issue of presentation than a lack of recognition of the complex interrelations between science and policy. Under each of these categories, importance is given to collaborative tools and mechanisms that account for the two-way directions and multiple iterations between science and policy. However, the absence of explicit reference and discussion about the implication of this complex interrelation between science and policy did not help showcase what constitutes one of the most important science-policy interface challenges. This could have consisted, for example, in discussing the order of presentation of the clusters in the text itself and, more fundamentally, in considering (reflexive) mechanisms (presented as cluster F "Define/refine the organizational business" in our conceptual framework) that could help in defining or refining these clusters over time within whatever mechanisms are put in place.

Besides these presentation issues, when it comes to content it is striking to note the almost complete absence of consideration for the diversity of information base and tools that need to be built to reflect the diversity of stakeholder engagement. The choice of putting emphasis on capacity development by opposition to capacity building may explain the lack of incorporation of potential mechanisms that could represent the various physical flow units under which a science-policy interface could be organized. This dimension is limited to data collection, which could be considered as a scientific bias for needs that could fall under this category.

Similarly, the only human capital aspect mentioned is training which, here again, remains quite limited and biased: it almost automatically implies training people from developing countries on methods and approaches developed in industrial countries. Exchange programs within countries or regions among different categories of stakeholders, personal and professional development, mentoring and coaching are not mentioned.

In any case, it seems that the potential discussion this document could have triggered on the type of mechanisms to put in place concretely did not really take place and was not reflected in the so-called Busan Outcome document that set out the main IPBES orientations.

This Busan Outcome recognized that IPBES would prioritize, support, and stimulate funding for capacity building and confirmed that capacity-building should be integrated "into all relevant aspects of its work according to priorities decided by the Plenary". We further discuss the consequences and importance of this operating principle in the next section.

Third phase (2011–2013): towards the establishment of IPBES Plenary meetings and the capacity building work program

During the 2011 and 2012 Plenary meetings to determine modalities and institutional arrangements for IPBES, much of the discussion focused on the potential role and mandate of IPBES. Members discussed what the Platform should focus on and how it could help integrate traditional knowledge and scientific knowledge for decision making and addressing biodiversity challenges. The second meeting, held in Panama in 2012, agreed to the resolution establishing an independent intergovernmental body to be known as IPBES. Attached to the resolution were the functions, operating principles and institutional arrangements of the Platform (including the establishment of the Bureau and the Multidisciplinary Expert Panel) and the rules of procedures to be used by the Platform.

Two Platform Plenaries took place in 2013 (in Bonn, Germany in January and Antalya, Turkey in December) and formally adopted the work program for 2014–2018 together with the conceptual framework. The creation of three task forces on capacity building, on knowledge and data and on indigenous and local knowledge systems was considered as part of the institutional arrangements needed to deliver the work program. Capacity building has an ubiquitous position within this work program. It is one component of the work program, together with knowledge generation, assessment and policy tools and methodologies, but because of the operating principle that states that capacity building should be integrated into all relevant aspects of its work according to priorities decided by the Plenary, it is also part of these three other activities.

In other words, as pointed out in the first working document of the first task force meeting (IPBES, 2014), there are two capacity building dimensions implicitly considered within IPBES. The first one is a general capacity building concept as applied to IPBES objectives as a science-policy platform for biodiversity and ecosystem services, while the second is a more inward-looking capacity building concept that is directly associated with supporting implementation of the work program at all appropriate levels.

Both dimensions are equally important – the second one, as forcefully requested by many groups during the IBPES negotiation process, is meant to ensure full participation of all parties in IPBES activities. The first, interestingly, establishes a dynamic and reflexive approach. This aspect is particularly important in such a bureaucratic context as there is a high risk that, in the long run, a circular rationale will be implemented whereby capacity building activities are meant for self-justification of IPBES itself regardless of potential changing needs and context. Establishing a mechanism to constantly refine the organizational business – through constant revalidation of the capacity building priorities and needs, performance measures and reporting, evaluation and feedback processes – is indeed critical to address changing needs, identify existing capacities, facilitate access to support, monitor capacity building delivery, assess effectiveness and provide feedback.

This reflexive dimension of capacity building has not been addressed in any previous discussion. Yet more than half of all the articles that define the capacity building task force mandate relate to this capacity building dimension.[7] We can assume that the importance given to this dimension is a reflection of the divergent views on the need to adopt a centralized or decentralized model to support capacity building.

The other elements of the mandate include human capital (fellowship exchange and training programs), institutional capital (to support building of the institutional capacity needed to implement the work program, particularly with respect to regional and subregional assessments), economic capital (to assist in addressing the prioritized capacity building needs agreed by the Plenary, drawing on resources made available through the Platform trust fund or provided through additional financial and in-kind support) and social capital (to liaise as necessary with the task force on knowledge and data and the task force on indigenous and local knowledge so as to ensure that capacity-building related to those issues is addressed in a consistent manner). The work program further specifies various activities to be conducted, such as "technical assistance, training workshops, fellowship and exchange programmes and support for the evolution of national, subregional and regional science-policy networks, platforms and centres of excellence, including where appropriate consideration of indigenous knowledge systems" (IPBES, 2013: 55).

Conclusion

Capacity building discussions have evolved over time in the IPBES establishment process. Starting with a broad collection of views that encompassed all aspects of capacity building, the first phases helped identify critical issues in relation to capacity building that structured the subsequent discussions. Two issues in particular were given considerable importance in the following years: the level of integration of capacity building activities in the work program of the future organization and the focus (or not) given to science-policy interface activities only (by opposition to a broader view of capacity building generally geared towards biodiversity and ecosystem services).

The first issue was crucial in helping reach a consensus among the various countries on the need for and establishment of IPBES. As for the second one, the final mandate given to IPBES and its task force clearly targets science-policy interface activities. It will however be interesting to apply the analysis grid suggested in this chapter to get a better sense of the levels of intervention (individual, collective, institutional, international) and types of capability (human capital, social capital, institutional capacity and economic capital) eventually targeted and how they could change over time.

Notes

1 Emphasis added
2 This could include bilateral donors as well as international organizations active on capacity building for biodiversity (e.g. the Global Environment Facility as a financing instrument and the Food and Agricultural Organization of the United Nations, the United Nations Development Programme, the United Nations Educational, Scientific and Cultural Organization, the United Nations Environment Programme, the World Bank and regional development banks).
3 Annex II of Decision IPBES-2/5 states that:

> the responsibilities of the task force are as follows: a) To develop modalities for identifying, monitoring and evaluating capacity-building needs relating to the Platform's mandate and programme of work, and promote their implementation in a consistent and comparative manner; b) To propose a process for systematic national self-assessment of capacity needs in the context of the Platform, when requested by governments, working with the Secretariat to implement such a process if and when agreed; c) To provide a draft list of priority capacity-building needs and an indication of associated financing gaps and available sources of funding; d) To periodically analyse the extent to which priority capacity-building needs identified by the Platform have been addressed and the role that the Platform has played in that process and to identify gaps and recommend ways in which such gaps could be addressed; e) To support the organization of the forum with conventional and potential sources of funding, in giving advice on the agenda and format of the meeting, participation, and how identified capacity-building needs and opportunities should be presented; f) To advise on the implementation of a 'matchmaking' facility to help to match available technical and financial resources with priority capacity-building needs, seeking and taking advice from the forum as appropriate; g) To propose means that could be developed for effectively integrating identified capacity building needs into the policies and programmes of development assistance processes, seeking advice from the forum as appropriate; h) To develop a proposal for fellowship exchange and training programmes; i) To support the building of the institutional capacity needed to implement the work programme, particularly with respect to regional and subregional assessments; j) To assist in addressing the prioritized capacity-building needs agreed by the Plenary, drawing on resources made available through the Platform's trust fund or provided through additional financial and in-kind support; k) To liaise as necessary with the task force on knowledge and data and the task force on indigenous and local knowledge so as to ensure that capacity-building related to those issues is addressed in a consistent manner.
>
> (IPBES, 2013: 65)

4 International Institute for Sustainable Development: www.iisd.ca.

5 COP 5 in 2000 in Nairobi "recognized that there is a need to improve the quality of scientific, technical and technological advice provided to the COP, and to undertake sound scientific and technical assessments on issues critical for implementation of the Convention. It requested SBSTTA to continue to improve the way it conducts its work (decision V/20 III, paragraphs 25, 26), and asked SBSTTA to identify and develop methods to undertake or participate in scientific assessments, and to identify and regularly update assessment priorities and information needs (decision V/20 III, paragraph 29)" (www.cbd.int/convention/sbstta-back.shtml).
6 UNEP, 2010b
7 See note 3, points a) to d).

References

Babin, D., ThibonM., LarigauderieA., GuinardS., MonfredaC. & BrelsS. 2008. *Strengthening the Science-policy Interface on Biodiversity. Results of the Consultative Process towards an IMoSEB*. IMoSEB/IFB/CIRAD.

Beck, S. 2010. Moving Beyond the Linear Model of Expertise? IPCC and the Test of Adaption. *Regional Environmental Change*, 11(2), 297–306.

Brahy, N. & Louafi, S. 2004. La convention sur la diversité biologique à la croisée de quatre discours. *Les rapports de l'IDDRI*, n°3.

Brand, U. & Görg, C. 2013. Regimes in Global Environmental Governance and the Internationalization of the State: The Case of Biodiversity Politics. *International Journal of Social Science Studies*, 1(1): 110–122.

Duraiappah, A.K. and Rogers, D. 2011. The Intergovernmental Platform on Biodiversity and Ecosystem Services: Opportunities for the Social Sciences. *European Journal of Social Science*, 24(3): 217–225.

Franks, T. 1999. Capacity Building and Institutional Development: Reflections on Water. *Public Administration and Development*, 19, 51–61.

IPBES 2013. Report of the Second Session of the Plenary of the Intergovernmental Science-Policy Platform on Biodiversity and Ecosystem Services. Second Session Antalya, Turkey, 9–14 December 2013, IPBES/2/17.

IPBES 2014. Discussion Paper for the First Meeting of the IPBES Task Force on Capacity Building. First Meeting, 21–23 May 2014, Trondheim, Norway.

Kleine, M. 2009. Capacity Building for Effective Work at the Interface of Forest Science and Forest Policy. *Mountain Research and Development*, 34.

Koetz, T., Bridgewater, P., van den Hove, S., Siebenhüner, B. 2008. The Role of the Subsidiary Body on Scientific, Technical and Technological Advice to the Convention on Biological Diversity as Science–Policy Interface. *Environmental Science & Policy*, 11: 505–516.

Koetz, T., Farrell, K.N. & Bridgewater, P. 2012. Building Better Science-Policy Interfaces for International Environmental Governance: Assessing Potential within the Intergovernmental Platform for Biodiversity and Ecosystem Services. *International Environmental Agreements*, 12(1): 1–21.

Le Prestre, P. (ed.) 2002. Governing Global Biodiversity. *The Evolution and Implementation of the Convention on Biological Diversity*. Brookfield: Ashgate.

Miller, C. 2001. Hybrid Management: Boundary Organizations, Science Policy, and Environmental Governance in the Climate Regime. *Science, Technology & Human Values*, 26(4), 478–500.

Miller, C. & Erickson, P. 2006. The Politics of Bridging Scales and Epistemologies: Science and Democracy in Global Environmental Governance. *Bridging Scales and Knowledge Systems: Concepts and Applications in Ecosystem Assessments.* Washington: Island Press.

MooreS.A., SevernR.C. & MillarR. 2006. A Conceptual Model of Community Capacity for Biodiversity Conservation Outcomes. *Geographical Research*, 44(4): 361–371.

Morin, J.-F. & Orsini, A. 2014. Policy Coherency and Regime Complexes: The Case of Genetic Resources. *Review of International Studies*, 40(2): 303–324.

Pielke, R.A. 2007. *The Honest Broker. Making Sense of Science in Policy and Politics.* Cambridge: Cambridge University Press.

Potter, C. & BroughR. 2004. Systemic Capacity Building: A Hierarchy of Needs. *Health Policy and Planning*, 19(5): 336–345.

Raustiala, K. & Victor, D.G. 2004. The Regime Complex for Plant Genetic Resources. *International Organization*, 58.

Reid, W. V., Berkes, F., Wilbanks, T. J. & Capistrano, D. (eds.) 2006. *Bridging Scales and Knowledge Systems: Concepts and Applications in Ecosystem Assessments.* Washington: Island Press.

Robins, L. 2008. Making Capacity Building Meaningful: A Framework for Strategic Action. *Environmental Management*, 42(5): 833–846.

Rosendal, K.G. 2001. Impacts of Overlapping International Regimes: The Case of Biodiversity. *Global Governance*, 7: 95–117.

Schacter, M. 2000. Capacity Building: A New Way of Doing Business for Development Assistance Organizations. Policy Brief No. 6. Ottawa: Institute on Governance.

Swanson, T. 1999. Why is there a Biodiversity Convention? The International Interest in Centralized Development Planning. *International Affairs*, 75(2): 307–331.

Swanson, T. & Groom, B. 2012. Regulating Global Biodiversity: What is the Problem? *Oxford Review of Economic Policy*, 28(1), 114–138.

UNEP 2005. Bali Strategic Plan for Technology Support and Capacity-building. UNEP/GC23/6/Add.1.

UNEP 2009. Report of the Second Ad Hoc Intergovernmental and Multi-stakeholder Meeting on an Intergovernmental Science-policy Platform on Biodiversity and Ecosystem Services. Nairobi, 5–9 October 2009, UNEP/IPBES/2/4/Rev.1.

UNEP 2010a. Report of the Third Ad Hoc Intergovernmental and Multi-stakeholder Meeting on an Intergovernmental Science-policy Platform on Biodiversity and Ecosystem Services. Busan, Republic of Korea, 7–11 June 2010, UNEP/IPBES/3/3.

UNEP 2010b. Analysis of Capacity Development for Biodiversity and Ecosystem Services. Note by the Secretariat, UNEP/IPBES/3/INF/3.

UNEP 2012. Report of the Second Session of the Plenary Meeting to Determine Modalities and Institutional Arrangements for an Intergovernmental Science-policy Platform on Biodiversity and Ecosystem Services. Panama City, 16–21 April 2012, UNEP/IPBES.MI/2/9.

Vadrot, A. 2013. Understanding the Establishment of the Intergovernmental Platform for Biodiversity and Ecosystem Services (IPBES): Epistemic Selectivities in International Biodiversity Politics. PhD Dissertation, April 2013, University of Vienna.

Van den Hove, S. & Chabason, L. 2009. The Debate on an Intergovernmental Science-Policy Platform on Biodiversity and Ecosystem Services (IPBES): Exploring Gaps and Needs. *IDDRI Working Papers*, 1. Paris: Institute for Sustainable Development and International Relations (IDDRI).

10 Indigenous challenges under IPBES

Embracing indigenous knowledge and beyond

Claudio Chiarolla[1] *and Annalisa Savaresi*[2]

Introduction

Scientific literature has emphasized indigenous and local knowledge (ILK) systems' critical contribution to our understanding of biodiversity and ecosystem services, their interactions with – and effects on – a changing climate (Sutherland et al., 2014; Tengö et al., 2014). Hence, a fundamental operating principle of IPBES requires that the Platform "recognize(s) and respect(s) the contribution of indigenous and local knowledge to the conservation and sustainable use of biodiversity and ecosystems."[3]

The matter of what ILK is has been long debated.[4] IPBES initially used a working definition[5] describing ILK as "a cumulative body of knowledge, practice and belief, evolving by adaptive processes and handed down through generations by cultural transmission, about the relationship of living beings (including humans) with one another and with their environment" (Berkes, 2012).[6] Noting that several different terms – such as indigenous, local or traditional knowledge, traditional ecological/environmental knowledge, farmers' or fishers' knowledge, ethnoscience, indigenous science and folk science – are utilized in the literature and in practice in reference to ILK,[7] IPBES has decided to use these terms interchangeably.[8] From the perspective of their legal protection, nevertheless, there is some scope to distinguish between various forms of ILK (Savaresi, 2016). As this chapter will show, in fact, while some international legal instruments and processes focus on specific forms of empirical and resource management knowledge, others concentrate more on the overarching system of beliefs, worldviews and institutions that support the continued production of ILK. These instruments and processes are characterized by different regulatory approaches. While some concentrate on the promotion and diffusion of ILK for non-commercial purposes, others specifically focus on the commercial potential of ILK and tend to design regimes of "enclosure" by restricting access to ILK. In spite of their specificities, these bodies of norms all attempt to address core regulatory questions, namely: the need to identify the characterizing features of ILK that fall within their scope, as well as of relevant ILK holders; and the need to simultaneously protect ILK and promote its deployment and development.

This chapter provides a reflection on how international law addresses the regulatory challenges associated with the protection and the promotion of ILK, and on what lessons can be learnt to address ILK in the IPBES context. It considers the role of ILK systems in the light of various relevant instruments and processes, as well as their potential shortcomings. It also considers the newly established participatory framework under the so-called IPBES Stakeholder Engagement Strategy, with a particular focus on indigenous peoples and local communities' (IPLCs) representation challenges. The chapter concludes by highlighting risks associated with the work of IPBES, but also opportunities for strengthening global protection of the rights of ILK holders and biocultural diversity.

ILK in international law

In spite of the fact that numerous international instruments deal with ILK, no universally agreed definition of ILK exists in international law. Intergovernmental efforts to set intellectual property standards in the framework of the World Intellectual Property Organization (WIPO) since 2009[9] have authoritatively described ILK as "the intellectual and intangible cultural heritage, practices and knowledge systems of traditional communities, including indigenous and local communities."[10] This term encompasses heterogeneous forms of empirical and resource management knowledge, including knowhow, practices, skills and innovations that can be applied in a wide variety of contexts, such as agriculture, healthcare, biodiversity conservation, natural resource management, genetic resources, architecture and construction technologies.[11] This composite list of potential applications gives a flavor of the diverse types of ILK than can be protected as intellectual property.

Not all types of ILK have potential for commercial application and commodification, however. Instead, different types of ILK raise different regulatory challenges, and have warranted the adoption of different regulatory approaches in international law. This distinction can be better appreciated by comparing regulatory approaches to the protection and promotion of ecological knowledge, on the one hand, with those concerning the use of non-human genetic resources, on the other.

Traditional ecological knowledge has been described as "a system of classification, a set of empirical observations about the local environment, and a system of self-management that governs resource use."[12] The main regulatory objective associated with ecological knowledge concerns the recognition of the contribution of ILK holders to modern science; as well as ensuring that interactions with scientists and policy makers occur in a way that is both respectful and protective of ILK holders, including the right not to interact at all. Regulatory approaches to address this kind of need are largely based on "inventorying" techniques; thereby practices that are considered beneficial are collected and promoted. These instruments do, however, generally include requirements to ensure the provision of prior informed consent

(PIC) of ILK holders, as well as to ensure that ILK holders share in the advantages engendered by the use of their knowledge by third parties – so-called benefit-sharing. Yet there is a remarkable lack of clarity as to the precise scope of state obligations in this regard.

Conversely, specific legal developments have addressed the question of empirical traditional knowledge on non-human genetic resources.[13] The rationale for addressing this specific type of ILK as a separate and self-standing issue relates to the potential to use it for commercial purposes. In turn, this raises the question of how to recognize and remunerate the contribution of ILK holders to commercial enterprises, for instance, in the agriculture, pharmaceutical and cosmetic sectors, and how to address concerns related to so-called "bio-piracy."[14] The response to this challenge has been the adoption of rather sophisticated regimes of enclosure, largely based on forms of "contractualization" and "propertization", with a variety of approaches to PIC and benefit-sharing matters.

International instruments therefore take a variety of approaches to ILK, which are characterized by the nature of the ILK at hand, as well as by the regulatory objectives that are being pursued. As a result, international instruments dealing with ILK can be placed on a continuum, with enclosure on the one end of the spectrum and open access on the other. The same may be said in connection with bilateral, as opposed to multilateral, benefit-sharing measures. The following sections review the approach to the protection of ILK in international law, with a view to highlighting the obligations and practices that are relevant for the work of IPBES.

ILK concerning natural resources

A number of international environmental agreements have addressed the matter of ILK. The first recognition of the role of ILK in an international environmental agreement came with the 1992 Convention on Biological Diversity (CBD). CBD specifically includes the obligation for Parties to respect, preserve and maintain ILK that is "relevant for the conservation and sustainable use of biological diversity."[15] The notion of ILK under CBD encompasses both empirical knowledge on the use of non-human genetic resources and various forms of practical knowledge on natural resource management. The distinguishing factor of ILK within the scope of CBD is its relevance for the conservation and sustainable use of biodiversity. CBD singles out indigenous and local communities as ILK holders. While these two sets of knowledge holders enjoy different levels of protection in international law, for the purposes of CBD they are bundled together and enjoy the same rights (Savaresi, 2013). The scope of knowledge-holders covered under CBD, however, has been criticized for leaving out ILK that is not held by, and does not persist in, urbanized or Western societies (Dutfield, 2001).

The protection of ILK under CBD hinges on the notion of benefit-sharing and PIC. Nevertheless, CBD provisions in this regard are heavily qualified and leave a wide margin of discretion to state parties. While CBD provisions on ILK hardly express clear cut obligations (Dutfield, 2001), over the years their interpretation has been fleshed out in a complex body of soft law guidance on how PIC ought to be obtained, as well as on procedures for cultural, environmental and social impact assessments for activities concerning sacred sites, lands and waters traditionally occupied or used by indigenous and local communities. In this regard, the CBD Akwé: Kon Guidelines provide voluntary guidance on participation and involvement of indigenous and local communities in development exercises as well as the incorporation of ILK as part of environmental, social and cultural impact assessment processes,[16] whereas the Tkarihwaié:ri Code of Ethical Conduct includes principles and methods to ensure that ILK is equally valued and considered alongside science.[17] CBD Parties have furthermore initiated a process to develop international guidelines on PIC and benefit-sharing from the use of ILK.[18] CBD obligations concerning ILK are therefore rather well established, and subject to an ongoing process of consolidation. CBD negotiations on these issues have become a forum for representatives of ILK holders, and especially indigenous peoples' representatives, to engage in international policy making, in ways that go well beyond the standard practice under other multilateral environmental agreements (Morgera, Tsioumani & Buck, 2014). Yet, indigenous peoples have oftentimes expressed concern over the approach to PIC embedded in CBD, asking for its alignment with the understanding of how this term emerged in human rights law.

Specific legal developments have furthermore occurred under the biodiversity regime in relation to ILK on non-human genetic resources. The 2010 Nagoya Protocol to the CBD[19] was specifically adopted with the aim of tackling bio-piracy concerns associated with the use of genetic resources and related ILK. The Protocol's approach to the protection and promotion of ILK is twofold. On the one hand, the Protocol allows contracting parties to restrict the lawful use of ILK for research and development purposes to instances where it has been obtained with the PIC of knowledge holders, in accordance with national legislation.[20] On the other, it requires that benefits generated from the utilization of ILK be fairly and equitably shared with ILK holders.[21] In both cases, the rights of IPLCs are subject to national legislation and depend on the latter for their recognition. So while states should provide an enabling domestic legal, regulatory or policy framework for bilateral agreements between providers and users,[22] the substance of these arrangements is often left – at least in part – to their contractual discretion. Therefore, while the Nagoya Protocol has advanced the protection of ILK associated with genetic resources through a regime that enables enclosure, it has largely left the definition of the terms of such enclosure not only to countries' priorities and approaches but also to contractual arrangements between the parties. This approach is not free of risk, given the typically weaker contractual position of ILK holders vis-à-vis their commercial counterparts in ABS contracts as well as vis-à-vis other stakeholders when domestic priorities are determined in Access and Benefit Sharing legislation.

The Nagoya Protocol, moreover, does not address all non-human genetic resources. Plant genetic resources for food and agriculture are in fact addressed in a dedicated international instrument, which prevails over the regime designed in the Nagoya Protocol, under certain circumstances.[23] The 2001 International Treaty on Plant Genetic Resources for Food and Agriculture (ITPGR)[24] includes a rather qualified call for the protection of ILK relevant to plant genetic resources for food and agriculture in national legislation, framing it as a component of farmers' rights.[25] As in the Nagoya Protocol, ITPGR deals with empirical ILK. ITPGR, however, does not address all ILK holders, but only concentrates on farmers. Moreover, contrary to the Nagoya Protocol, ITPGR does not purport to establish a bilateral system to regulate access to and use of ILK. The reason for this is that by their own very nature plant genetic resources for food and agriculture require continued use and exchange, and are the result of developments that may not just be attributed to a single group or set of users. Therefore, ITPGR has introduced a multilateral system to facilitate the exchange of plant genetic resources and "associated available non-confidential descriptive information",[26] which does not require the provision of PIC on a case by case basis (Chiarolla, Louafi & Schloen, 2013). The Multilateral System also enables farmers to share in the benefits arising from the use of plant genetic resources,[27] by receiving funding to conserve crop diversity and adapt them to changing needs and demands.[28] Therefore, the matter of both PIC and benefit-sharing is addressed in rather different terms under ITPGR and the Nagoya Protocol.

A regime of enclosure is also being considered in the context of the World Intellectual Property Organization where, since 2000, an International Committee has been working towards the development of an international legal instrument for the protection of, *inter alia*, traditional knowledge. The regulatory approach here, however, clearly hinges on intellectual property law and on strategies to prevent, on the one hand, unauthorized use by third parties (positive protection) and, on the other, that third parties illegitimately acquire intellectual property rights over ILK (defensive protection). The forms of ILK that come into play under WIPO are therefore those with a potential to be subjected to intellectual property-related measures. Singling out instances where this may be the case is a controversial exercise, and limited progress has been made in elaborating a proprietary approach to traditional knowledge protection under WIPO.

Finally, yet another regulatory approach to ILK is embedded in the 1994 UN Convention to Combat Desertification (UNCCD).[29] Because of its subject matter, UNCCD covers a wide range of ecological and empirical knowledge that is relevant to address desertification. Using rather qualified terms, the Convention requires its Parties to exchange information on ILK, while ensuring adequate protection and providing appropriate returns from the derived benefits to the local populations concerned, on an equitable basis and on mutually agreed terms.[30]

Contrary to CBD and ITPGR, UNCCD generally refers to "local populations" and traditional knowledge "owners", thus possibly also including individuals in the scope of protection.[31] UNCCD does not stipulate the PIC requirement for ILK use. Rather, it requires Parties to prepare inventories of such knowledge and their potential uses "with the participation of local populations, and disseminate such information, where appropriate, in cooperation with relevant intergovernmental and non-governmental organizations."[32] UNCCD therefore relies on inventorying to promote and divulge the use of ILK, merely by requiring that access be in line with customary practices governing access to specific aspects of such knowledge.[33] Whether or not this can be read as a PIC requirement is a matter for interpretation. Furthermore, benefit-sharing seems only to come into play when some form of commercial exploitation is at stake.[34]

This overview shows that international environmental law instruments have taken differentiated approaches to ILK protection and to PIC use and benefit-sharing arrangements. The rationale for these different approaches largely relates to the nature of the ILK at stake, as well as the regulatory objectives that are being pursued. Where the objective is to promote the use of ILK, the international instruments reviewed here do not insist on the provision of specific benefit-sharing requirements. These requirements are instead centerpieces in instruments concerned with the potential commercialization of ILK.

ILK as cultural heritage and in human rights law

Beyond the realm of international environmental law, the matter of ILK has been addressed in the context of cultural heritage and human rights law. These two bodies of law, nevertheless, address ILK in rather different fashions.

ILK falls within the scope of the 2003 Intangible Cultural Heritage Convention, which includes the practices, knowledge and skills that "communities, groups and, in some cases individuals, recognize as part of their cultural heritage."[35] Therefore, even though not all forms of intangible cultural heritage are ILK, ILK can be regarded as an expression of cultural heritage. Yet, the UNESCO Convention does not establish a regime of enclosure. Instead, it relies on inventorying as a means to give international visibility to intangible cultural heritage, especially that in need of urgent safeguarding. The listed forms of ILK include, for example, practices and know-how concerning tree species, and the cultivation of mastic. Therefore, even though the UNESCO Convention does not specifically deal with ecological knowledge or knowledge related to genetic resources, it is possible that these forms of knowledge fall within its scope of protection. However, contrary to CBD, the UNESCO Convention does not require the provision of PIC, or benefit-sharing arrangements. Instead, it says that inventories of ILK present in Parties' territory be compiled with the "participation of communities."[36] The cultural heritage regime has received some criticism for this and much emphasis has been placed on its cross-fertilization with human rights law.[37]

ILK has also been dealt with in the context of human rights law and practice.[38] Human rights bodies have addressed the matter of indigenous peoples' cultural heritage out of concerns that international instruments dealing with ILK from a sectoral perspective are not adequately protecting indigenous peoples' rights (Wiessner, 2013). In this context, the need to complement processes dealing with ILK under WIPO and CBD with a human rights-based approach to the protection of indigenous peoples' cultural heritage has been underscored.[39]

More generally, the matter of the protection of ILK falls within the scope of the right to culture, which has been protected under both foundational international human rights treaties since 1966.[40] The right to culture encompasses ways of life, religion or belief systems, rites and ceremonies, methods of production or technology, as well as "customs and traditions through which *individuals, groups of individuals and communities* express their humanity and the meaning they give to their existence, and build their world view representing their encounter with the external forces affecting their lives."[41] This broad range of issues clearly also encompasses ILK.

The right to culture imposes upon states both negative obligations of non-interference with the exercise of cultural practices and access to cultural goods and services; as well as positive obligations associated with ensuring preconditions for participation, facilitation and promotion of cultural life, and access to and preservation of cultural goods.[42]

In interpreting states' obligations associated with the right to culture, human rights bodies have drawn attention to the need to respect and protect cultural heritage of all groups and communities in economic development and environmental policies and programs and avoid the adverse consequences of privatization of goods and services.[43] The case law of international and regional human rights bodies[44] has specifically recognized the fundamental role of culture in the preservation and advancement of the identity and collective rights of indigenous peoples, as well as of all individual rights of their members.[45] This translates into the right to act to ensure respect for indigenous peoples' right to maintain, control, protect and develop their ILK, as well as the manifestations of their sciences, technologies and cultures, including human and genetic resources, seeds, medicines, knowledge of the properties of fauna and flora, oral traditions, literature, designs, sports and traditional games, and visual and performing arts.[46] In this regard, a link has been drawn between the need to respect and protect ILK, and the protection from illegal or unjust exploitation of their lands, territories and resources by state entities or private or transnational enterprises and corporations.[47] This protection has been also linked with other human rights, such as the prohibition of discrimination[48] and the right to self-determination.[49] In this regard, international human rights law recognizes indigenous peoples' right to free PIC,[50] which, according to some, has become part of customary international law.[51]

Regardless of whether one shares such a view, the right to free PIC has been widely subsumed in the practice of development assistance. Human rights law also specifically recognizes indigenous peoples' right to benefit-sharing associated with the use of their traditional knowledge.[52]

Altogether the protection allocated to ILK under human rights law is a-specific and largely focuses on protecting the way of life and the circumstances that enable the creation and preservation of ILK. Thus, human rights law encompasses not only empirical and resource management knowledge, but also knowledge institutions and cosmologies. Contrary to the sectoral approaches in international environmental law instruments, therefore, human rights law addresses ILK in all-encompassing terms, and covering a wider range of traditional knowledge holders. It has done so by fleshing out a set of core safeguards for the protection of ILK which encompasses free PIC as well as benefit-sharing, at least in relation to indigenous peoples.

ILK in international climate law and policy making

Analogies are often drawn between IPBES and the Intergovernmental Panel on Climate Change (IPCC), the scientific body entrusted to periodically assess the magnitude, timing and potential environmental and socioeconomic impacts of climate change, as well as of possible response strategies.[53] Contrary to the biodiversity regime, however, the climate regime has so far paid limited attention to ILK.

IPCC first acknowledged the role of ILK as a means to assess climate change, as well as to develop adaptation and natural resource management strategies in 2007. As to the first, IPCC recognized that traditional ecological knowledge could be a "useful source" to detect environmental changes associated with increased concentrations of greenhouse gases in the atmosphere, and therefore work as a complement to other sources of scientific knowledge (IPCC, 2007). This was said to be the case particularly with changes observed by "communities living in environments vulnerable to early climate change impacts", such as the Arctic, high-altitude zones, desert margins and low-lying coastal areas. As to the second, IPCC has recognized the role of ILK as a means to plan and implement strategies to adapt to a changing climate. More specifically, IPCC has recently recommended the "mutual integration and co-production" of traditional and scientific knowledge to increase adaptive capacity and reduce vulnerability (Adger et al., 2014). Like IPBES, therefore, IPCC is confronted with the challenges of how to actually facilitate these outcomes in practice.

In this regard, some guidance may come from the international instruments and processes identified above. Indeed, various forms of ILK are potentially applicable to climate change adaptation, and in some instances ILK falling within the scope of other international environmental instruments may be relevant. In other cases, ILK falling within the scope of international instruments on the protection of cultural heritage and human rights may be at stake. The preponderant concern in the context of climate change, however, is not so much the protection of ILK, but rather the promotion of its use. Therefore, as seen in the context of UNCCD, the climate regime has increasingly emphasized the need of promoting inventories of relevant ILK. Yet, no specific legal obligations have been imposed upon United Nations Framework Convention on Climate Change (UNFCCC) Parties in this regard. So far, instead, Parties to the climate regime have merely been invited to underline the importance of ILK and practices for adaptation. The financial mechanism of the climate regime has been encouraged to enhance consideration of ILK and practices and their integration into adaptation planning and practices, as well as procedures for monitoring, evaluation and reporting.[54] These rather scant recommendations hardly impose any obligations on UNFCCC Parties.

Equally, IPCC has not created institutional pathways to systematically involve ILK holders in its work, along the lines of IPBES.[55] The application of ILK to climate change, nevertheless, raises a host of questions concerning the maintenance of lifestyles that enable the continued creation and transmission of ILK; securing interaction between modern science and ILK in a way that is respectful of the latter; and enabling ILK holders to participate in the design of climate change response measures. As seen earlier, these questions have already been considered, at least to some extent, in the context of extant international instruments and processes dealing with ILK.

The only area of the climate regime where the potential for cross-fertilization with these instruments and processes has been explicitly acknowledged so far is that of measures to incentivize developing countries' forest carbon storage capacity, commonly referred to with the acronym "REDD+".[56] In this context, UNFCCC Parties have undertaken some steps to recognize and address overlaps with States' obligations concerning ILK under human rights law and biodiversity law. Still, the matter of ILK in the climate regime raises particularly complex and multifaceted regulatory questions, whose answers are yet to be written. In this regard, important lessons can be drawn from the implementation of extant international instruments dealing with ILK and from the practice that is being developed in the context of IPBES (Savaresi, 2015).

Legal and institutional challenges arising from the recognition of the role of ILK under IPBES

The evolution of international law concerning ILK has been characterized by a tension between promoting access for the public good and the need to protect the culture, livelihood and identity of ILK holders, especially vulnerable groups, such as indigenous peoples. The international law instruments dealing with these separate, but inherently interrelated objectives are far from having reached maturity and have some shortcomings.

Obligations concerning ILK under international biodiversity law have engendered concerns over their adherence to human rights standards, especially concerning the protection of the rights of indigenous peoples. More generally, the design and implementation of regimes of exclusivity or enclosure for the protection of ILK has also been difficult, unpractical or ineffective.

Little systemic thinking has been done with regard to the promotion of ILK. In this regard, arguably the most detailed obligations come from international human rights law, and states' obligations in relation to the protection of the right to culture, as well as of cultural heritage. In addition, some voluntary guidance has been adopted in the context of CBD. These bodies of law, however, hardly address the contribution of ILK holders to the science-creation process. In this connection, IPBES is arguably breaking new grounds by navigating unchartered waters.

Challenges concerning the IPBES conceptual framework and ILK

The IPBES conceptual framework (CF) integrates ILK systems and aims to provide a common analytical basis for all IPBES stakeholders, including indigenous and local communities. It addresses the Platform's objective, functions and relevant operating principles.[57] IPBES CF aims to capture "the relationships between the natural world and humankind in only six main elements – nature, nature's benefits to people, anthropogenic assets, indirect drivers of change (such as institutions and governance systems), direct drivers of change, and good quality of life" (Díaz et al., 2015).

IPBES CF considers ILK as an anthropogenic asset, underscoring that "a good life is achieved by a co-production of benefits between nature and societies."[58] Under this conceptual horizon, customary norms from ILK systems may be viewed as "institutions and governance systems, and informal social norms and rules." In turn, Diaz et al. suggest that these institutions and governance systems may be regarded as indirect drivers of change, i.e. "external factors that affect nature, anthropogenic assets, nature's benefits to people and a good quality of life." As indirect drivers, customary norms that govern ILK systems may impact or underpin the "root causes of the direct anthropogenic drivers that affect nature" (Díaz et al., 2015). It is therefore evident that legal norms aimed at protecting these institutions and governance systems are of great importance in the work of IPBES.

Yet, IPBES CF highlights the potentially ambivalent role of ILK systems. On the one hand, they may positively contribute to achieving a "good quality of life" (e.g. through traditional agricultural knowledge concerning variety selection and agricultural practices or traditional medicinal knowledge). On the other, if resource use is not properly regulated, traditional normative systems associated with ILK – as any other normative system – may also promote practices that are detrimental to biodiversity (e.g. through their indirect impact on direct anthropogenic drivers such as the overexploitation of wild species).

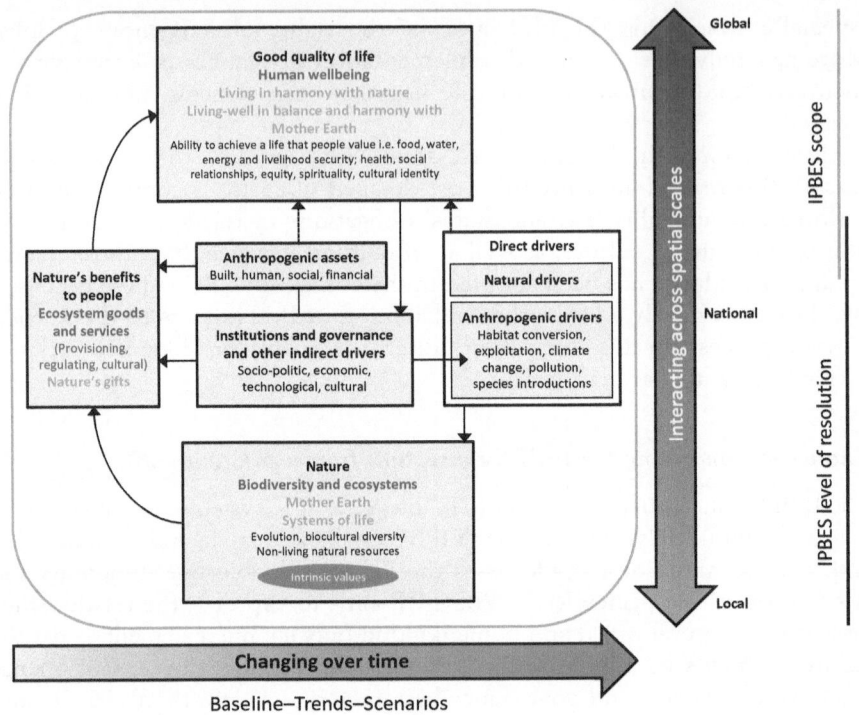

Figure 10.1 IPBES conceptual framework
Source: IPBES/2/17, p. 40.

While the evaluation of the impact of different management practices on biodiversity could be challenging *per se*, ILK holders may be reluctant to provide information about their management practices in assessment processes over which they exercise neither ownership nor control and that are undertaken in accordance with methodologies and criteria they are not familiar with. In the context of climate change adaptation, for instance, grassroots responses to tackle these concerns have emerged with the launch of the Indigenous Peoples' Bio-cultural Climate Change Assessment Initiative.[59] It follows that a key challenge for IPBES will be to provide appropriate incentives and safeguards to ILK holders for sharing knowledge relevant to the assessment and other functions of the Platform, while preserving the integrity of the link with value systems from which such knowledge is extrapolated.

Another important challenge relates to the implication of using ILK, alongside scientific knowledge, for the production of various Platform deliverables, particularly with regard to its implications for the definition of the most appropriate assessment scale. In light of the prevalent relevance of ILK for local and national assessments, the relatively liberal inclusion of references to ILK values, systems and stakeholders in the work program and conceptual framework of the Platform – which was an important requirement for several ALBA countries,[60] among others – has been criticized for its potential to jeopardize the work of IPBES (Soberón & Townsend Peterson, 2015). This is because the focus on ILK systems might steer attention away from the main Platform mandate regarding global and regional scale assessments vis-à-vis local and national assessments, whose responsibility should rest instead with national governments and institutions.

Challenges concerning the task force on ILK systems

IPBES members have agreed to establish a task force on indigenous and local knowledge systems, and another on knowledge and data. The former task force will establish, *inter alia*, a participatory mechanism to support indigenous and local knowledge system contributions to the Platform deliverables, facilitate linkages between IPLCs and scientists, and strengthen indigenous peoples' participation in the Platform work. This is indeed an important institutional innovation as compared to IPCC – and all other treaty mechanisms considered in this chapter – since it institutionalizes modalities and procedures devoted to enabling IPLC participation as an "ordinary" feature of IPBES.

The IPBES task force on ILK systems consists of two Bureau members, three members and one back-up member of the Multidisciplinary Expert Panel, between them covering the five United Nations regions, and up to 20 additional experts on indigenous and local knowledge systems.[61] As suggested by the above described composition, the possible identification, selection and nomination of legitimate IPLC representatives as task force members is not a guaranteed outcome. Indeed, being an expert on ILK systems is not the same as being a legitimate representative of knowledge holders within such systems. However, the representation of ILK holders amongst the task force members is possible under the current rules. This representation is also essential to minimize future criticism in light of the relatively inclusive practices concerning IPLC participation in the CBD Traditional Knowledge work program.[62]

Five work streams were agreed at the first task force meeting in 2014: to support the establishment of approaches and procedures for working with indigenous and local knowledge (ILK) systems; to undertake piloting of preliminary approaches and procedures for working with ILK systems in assessments; to oversee the convening of global dialogue workshops and the development of regional case studies; to support the establishment of a participatory mechanism for ILK systems to facilitate linkages between IPLCs and scientists; and the establishment of a roster and network of experts in ILK to support the Platform work.

With regard to the first work stream, while the task force is striving to elaborate the final approaches and procedures for working with ILK (with a view to their adoption by the Plenary), interim approaches and procedures were also developed in order to integrate ILK into ongoing regional and thematic assessments,[63] such as the one on pollination and pollinators associated with food production.[64] The proposed participatory mechanism should be responsible for promoting participation of IPLCs at different scales under IPBES, including the strengthening of national participatory mechanisms and the bridging role of academic institutions.

Task force on knowledge and data and ILK

The task force on knowledge and data has developed a Knowledge, Information and Data (KID) Plan as part of a broader strategy that aims to guide the work of the task force, "providing a context for other deliverables involving knowledge, information and data." Both the KID Plan as well as the draft KID strategy include provisions on handling ILK in IPBES assessments and on the "rights and attribution" to knowledge holders.

The recently adopted "Data and Information Management Plan" recognizes, *inter alia*, the following principles for managing knowledge, information and data in the Platform:

The custodians of data and knowledge essential to the Platform work programme are many and diverse, and the programme can only be delivered through collaboration. Consequently, the plan will [*inter alia*] recognize the needs and interests of custodians of data and knowledge, such as access rights and intellectual property rights, in particular the need to respect information provided by and the knowledge of indigenous peoples and local communities, which includes, as appropriate, consideration of seeking prior informed consent or approval and the involvement of indigenous peoples and local communities, who are holders of such information and knowledge, and the sharing of benefits accrued from such information and knowledge.[65]

Similarly, the proposed KID strategy to be developed in 2015 includes a specific section on "rights and attribution." In particular, this section indicates the intention to develop guidelines for "handling the rights of knowledge holders, including issues of transparency, acknowledgement, recognition, intellectual property, access, and respect for indigenous knowledge."[66] Hence, the above operative principles are a clear reference to norms and obligations concerning indigenous and local communities' rights, benefit-sharing and free PIC procedures that are directly derived from international law. Hence, it is acknowledged that ILK treatment under the Platform is to be in line with international legal instruments such as the Nagoya Protocol and UNDRIP.

However, as illustrated in the second section of this chapter, several gaps still exist concerning the protection of ILK in international law. For instance, from an intellectual property standpoint, no international instruments provide minimum standards for ILK protection, and no mutual recognition of rights or normative harmonization of any kind has been agreed. Lack of agreement at the international level regarding the need for (and the most appropriate scope of) IP protection for traditional knowledge has thus far jeopardized efforts to establish new international IP standards at WIPO.

In addition, the body of international norms that regulate the possibility of granting entitlements to traditional knowledge holders under the Nagoya Protocol provides a high degree of flexibility and implementation options regarding its access and benefit-sharing obligations. Such flexibility, however, makes it difficult to derive globally applicable rules without reference to specific applicable laws – if any – of "provider" countries.

As a result of these gaps in the law, at present there is no set of rules to ensure a level playing field for exchanges between scientific knowledge and ILK. On the one hand, modern scientific knowledge, including most IPBES outputs and deliverables, can be subject to proprietary rights, primarily under copyright law. On the other, ILK is often implicitly treated as a basic research input unworthy of protection as such. Obviously, the risks that may derive from making ILK widely available, such as the non-attribution of authorship, can be minimized when assessments are undertaken directly by the affected communities, since copyright will be vested with the knowledge providers and evidence of the source will be implicit. While awareness of the above issue is increasing among indigenous peoples and local communities, it does not seem to have constituted an obstacle to mobilizing available ILK in the climate change assessment setting. In this regard, a recent thematic paper highlights that "observations and assessments by indigenous peoples have remained relatively inaccessible to the IPCC process mostly due to language and socio-cultural barriers. Thus, for the most part, indigenous and marginalized peoples' knowledge that appears in grey literature … or that is made available through non-written media has remained outside the scope of IPCC assessments."[67]

Besides the delivery of assessment-related capacity development to IPLCs, in many cases, a legal level playing field in this area could be beneficial. Some degree of harmonization and possibly agreed minimum standards for ILK protection would help address outstanding questions such as: Who are the legitimate owners/holders of ILK, including that which is used in the context of IPBES work? Who can legitimately grant permission to use it or set limitations to its use (e.g. regarding culturally appropriate use)? What is the role of customary law as a condition of access to ILK and how can that role be recognized under IPBES? How, when and who should define and share benefits arising from the envisaged IPLC participation in IPBES (e.g. through capacity building)? Should the holders/custodians of ILK vest any rights to knowledge products that incorporate or make use of their ILK?

At present, international obligations associated with the rights of indigenous peoples and, in particular, the human right to culture, as well as with the Nagoya Protocol and bio-cultural community protocols[68] are the most important normative references to be taken into account in the IPBES context. Indeed, if IPBES manages to provide further operational guidance on these questions in a way that balances the rights and interests of IPLCs with the needs of the scientific community, it could certainly help improve the understanding of how the rights of ILK holders can be implemented in the context of intergovernmental assessments and in other law and policy-making fora.

Stakeholder engagement strategy and IPLC participation

A key objective of the IPBES stakeholder engagement strategy is to attract "scientists, such as assessment experts, and other knowledge holders from citizen science initiatives and indigenous peoples and local communities to contribute to the Platform deliverables."[69] However, a critical concern voiced by IPLCs is that they have not been successful in gaining recognition as a distinct group of stakeholders, but only formally participate in IPBES activities as one group amongst many diverse stakeholder groups, which also include academics, NGOs, industry, scientific and research bodies and others.[70] In particular, indigenous representatives have stressed that they intend to continue to interact with IPBES in their own capacity as knowledge and rights holders rather than as simple stakeholders.[71] In particular, they claim that they should be treated as full partners of IPBES, in line with the positions often expressed in CBD (and in other fora), where recognition of their distinct identities[72] is considered as a necessary precondition for their full and effective participation in relevant decision and policy-making processes.

While the IPBES strategy does not identify the specific incentive mechanisms that will need to be mobilized in order to engage with ILK holders, it is clear that its three task forces will offer important opportunities for stakeholder engagement, as their work will heavily rely on collaborations with various partners in the field of capacity building, indigenous and local knowledge, and knowledge and data. Furthermore, with regard to outreach activities, the IPBES implementation plan recognizes that language barriers and other constraints, such as limited funding or limited internet access, are critical challenges in pursuing stakeholder engagement, particularly with ILK holders.

Especially for IPLCs already working on biodiversity assessments and indicators at the community and national levels, funding availability and time constraints are key factors shaping their concrete opportunities to participate in regional and international assessments. The full and effective participation of IPLCs in IPBES work will only be possible if both the implementation of the stakeholder engagement strategy and the delivery of assessment-related capacity development manage to reduce these constraints. With respect to capacity development, particular attention should be paid to avoiding a one-size-fits-all approach that would not do justice to the different conditions under which ILK is preserved and transmitted, and to the diverse engagement motivations of different knowledge holders. In this regard, a foundational point of departure of capacity-building activities should be the recognition and respect of relevant customary norms and related bio-cultural community protocols, whenever they are available.

Conclusions

Over the years, international instruments concerning a range of matters, such as biodiversity, agriculture, cultural heritage and human rights, have addressed the ILK issue. Some of them have focused on means to protect ILK by allowing countries to provide property-like entitlements to groups or individual holders and set restrictions on the ways in which such knowledge may be used by third parties. Others, instead, have placed greater emphasis on establishing means by which ILK ought to be promoted, for instance, by facilitating its divulgation though inventorying processes.

One area of crucial legal development has been that concerning the relation between the consent of indigenous peoples and fair and equitable benefit-sharing as a means to recognize, support and reward ILK holders for their contribution to the pursuit of global public goods. These developments, nevertheless, have only marginally influenced the climate regime, where measures to address ILK and its contribution to climate change adaptation, assessment and observation are yet to be adopted. In this regard, important lessons may be drawn from biodiversity law and the evolving experience of IPBES. For instance, many of the normative and aspirational principles developed within the remit of biodiversity and human rights law have been integrated in the work of IPBES. The operating principles and institutional arrangements of IPBES are formally respectful of fundamental requirements concerning IPLC participation and involvement in the Platform work and of the need to seek and obtain free prior informed consent.

It is nevertheless true that the practical implementation of these requirements is not straightforward in many cases. It depends on the extent to which capacity building under IPBES will match the needs of IPLCs to enable their participation and on the provision of adequate funding. It also depends on whether specific IPLCs (or groups of individuals within such communities) can be identified as the holders of relevant ILK. It depends on whether authoritative, representative and legitimate governance structures are available to discuss and provide their contributions to the science-policy interface. Moreover, it depends on national legislation and on the extent to which the latter recognizes and respects the interests, prerogatives and rights of IPLCs under different circumstances. Finally, it depends on governments' political will to allow indigenous voices to be heard in national, regional and international assessments on biodiversity.

In its treatment of ILK, IPBES is navigating unchartered waters and facing unprecedented challenges. Yet these unchartered waters are not being navigated without the compass of international instruments and processes dealing with ILK. In taking the lead with implementing international guidance on ILK, IPBES is engaging in an exercise of potentially momentous consequences. The outcomes of this exercise will be crucial for the future of knowledge co-creation and management in relation to biodiversity and ecosystems. They are likely to have profound implications also for other international processes dealing with ILK, most saliently in relation to climate change. The stakes could not be higher as a new frontier is open to strengthen the global protection of indigenous and local knowledge systems and bio-cultural diversity.

Notes

1 Legal Officer, World Intellectual Property Organization (WIPO) Traditional Knowledge Division. The views expressed in this chapter are the authors' own and do not necessarily reflect the views of WIPO or any of its member states. The research for this chapter was undertaken when the author was Research Fellow at the Institute for Sustainable Development and International Relations (IDDRI) in Paris.
2 Research and Teaching Fellow in Global Environmental Law, University of Edinburgh Law School, United Kingdom.
3 IPBES. 2012. Functions, operating principles and institutional arrangements of IPBES. http://www.ipbes.net/sites/default/files/downloads/Functions%20operating %20principles%20and%20institutional%20arrangements%20of%20IPBES_2012. pdf. Accessed 13 June 2016.
4 For consistency, we will use only the term ILK throughout this chapter. However, ILK is often also referred to as "traditional knowledge" both in the literature and in relevant international legal instruments. Hence, in principle, we have assumed that these terms may be used interchangeably.
5 IPBES. 2013. Consideration of initial elements: recognizing indigenous and local knowledge and building synergies with science, IPBES/1/INF/5, Annex A. http:// www.ipbes.net/sites/default/files/downloads/IPBES_1_INF_5.pdf. Accessed 13 June 2016.
6 IPBES. 2013. Consideration of initial elements: recognizing indigenous and local knowledge and building synergies with science, para 3.
7 IPBES. 2013. Consideration of initial elements: recognizing indigenous and local knowledge and building synergies with science, Annex A.
8 Ibid.
9 WIPO, The WIPO Intergovernmental Committee on Intellectual Property and Genetic Resources, Traditional Knowledge and Folklore, Background Brief N°2. www.wipo.int/export/sites/www/tk/en/resources/pdf/tk_brief2.pdf. Accessed 13 June 2016.
10 Intergovernmental Committee on Intellectual Property and Genetic Resources, Traditional Knowledge and Folklore (IGC), Glossary of Key Terms Related to Intellectual Property and Genetic Resources, Traditional Knowledge and Traditional Cultural Expression (2012) [hereafter "Glossary of Key Terms"], p. 44. http://www.wipo.int/meetings/en/doc_details.jsp?doc_id=192978.   Accessed   13 June 2016.
11 Intergovernmental Committee on Intellectual Property and Genetic Resources, Traditional Knowledge and Folklore (IGC), The Protection of Traditional Knowledge: Draft Articles (2014) WIPO/GRTKF/IC/28/5, at 5. http://www.wipo. int/meetings/en/doc_details.jsp?doc_id=276361. Accessed 13 June 2016.
12 Glossary of Key Terms, p. 42.
13 Convention on Biological Diversity (CBD). (Rio de Janeiro, 5 June 1992, in force 29 December 1993) Decision II/11, Access to genetic resources, UNEP/CBD/COP/ 2/19, para 2.
14 The term "bio-piracy" is used to refer either to the unauthorized commercial use of biological resources and/or associated ILK from developing countries, or to the patenting of inventions based on such knowledge or resources without compensation.

15 CBD, 1760 UNTS 79, Article 8(j).
16 CBD Decision VII/16C (2004), Annex, Akwé: Kon voluntary guidelines for the conduct of cultural, environmental and social impact assessments regarding sacred sites and lands and waters traditionally occupied or used by indigenous and local communities, which focus on how to take into account traditional knowledge, innovations and practices as part of the impact-assessment processes and promote the use of appropriate technologies, at 59–60.
17 CBD Decision X/42 (2010), Annex, Tkarihwaié:ri Code of Ethical Conduct to ensure respect for the cultural and intellectual heritage of indigenous and local communities relevant to the conservation and sustainable use of biological diversity, which provides for prior informed consent and/or approval and involvement of traditional knowledge holders.
18 CBD Decision XII/12D (2014), at 2.
19 Nagoya Protocol on Access to Genetic Resources and the Fair and Equitable Sharing of Benefits Arising from their Utilization to the Convention on Biological Diversity (adopted at Nagoya, 29 October 2010, in force 12 October 2014).
20 Nagoya Protocol, Article 7.
21 Nagoya Protocol, Article 5(5).
22 Nagoya Protocol, Article 5(5).
23 Morgera, Tsioumani & Buck (2014) at 103.
24 International Treaty on Plant Genetic Resources for Food and Agriculture (ITPGR). (Rome, 3 November 2001, in force 29 June 2004) 2400 UNTS 303.
25 ITPGR, Article 9(2).
26 ITPGR, Article 12.3(c).
27 ITPGR, Articles 10 and 13.
28 ITPGR Article 18(4).
29 UN Convention to Combat Desertification in Countries Experiencing Serious Drought and/or Desertification, Particularly in Africa (UNCCD). (Paris, 14 October 1994, in force 26 December 1996) 1954 UNTS 3.
30 UNCCD, Article 16(g).
31 UNCCD, Articles 16(g) and 17(1)(c).
32 UNCCD, Article 18(2)(a).
33 UNESCO Convention on Intangible Cultural Heritage, Article 13(d)(ii).
34 UNCCD, Articles 17(1)(c) and 18(2)(b).
35 Convention for the Safeguarding of the Intangible Cultural Heritage (17 October 2003, in force 20 April 2006) 2368 UNTS 1, Article 2.
36 UNESCO Convention on Intangible Cultural Heritage Articles 11(b) and 12(1).
37 Human Rights Council, Promotion and protection of the rights of indigenous peoples with respect to their cultural heritage. Study by the Expert Mechanism on the Rights of Indigenous Peoples, 2015, A/HRC/EMRIP/2015/2, Annex.
38 UN Draft Principles and Guidelines on the Protection of the Heritage of Indigenous People, elaborated by the Special Rapporteur of the Sub-Commission on the Promotion and Protection of Human Rights, E/CN.4/Sub.2/1995/26, and approved in 2000 by the United Nations Sub-Commission on the Promotion and Protection of Human Rights.
39 UN Commission on Human Rights, Review of the draft principles and guidelines on the heritage of indigenous peoples. Expanded working paper submitted by Yozo Yokota and the Saami Council on the substantive proposals on the draft principles and guidelines on the heritage of indigenous peoples, E/CN.4/Sub.2/ AC.4/2005/3, at 7–11.
40 International Covenant on Civil and Political Rights (ICCPR) (New York, 16 December 1966, in force 23 March 1976), 999 UNTS 171, Article 27; and International Covenant on Economic, Social and Cultural Rights (ICESCR) (New York, 16 December 1966, 3 January 1976) 993 UNTS 3, Article 15.

41 UN Committee on Economic, Social and Cultural Rights, General comment No. 21, Right of everyone to take part in cultural life (2009) E/C.12/GC/21 [hereafter "General comment No. 21"], at 13. Emphasis added.
42 Ibid., at 6.
43 General comment No. 21, para 50.
44 See for example IACrtHR, *Case of the Moiwana Community v Suriname*, Series C No. 124, at 86(9); and ACHPR, *Centre for Minority Rights Development (Kenya) and Minority Rights Group International on behalf of Endorois Welfare Council v Kenya*, Communication No. 276/2003, para 241.
45 United Nations Declaration of the Rights of Indigenous Peoples (UNDRIP), March 2008; in particular Articles 5, 8, and 10–13, and 31. See also International Labour Organization (ILO) Convention No. 169, in particular Articles 2, 5, 7, 8, and 13–15. See also General comment No. 21, para 36.
46 ILO Convention No. 169, Articles 5 and 31. See also UNDRIP, Articles 11–13.
47 General comment No. 21, at 51 (c, d) and Human Rights Council, Promotion and protection of the rights of indigenous peoples with respect to their cultural heritage. Study by the Expert Mechanism on the Rights of Indigenous Peoples, 2015, A/HRC/EMRIP/2015/2, para 59.
48 ICCPR, Article 26.
49 ICCPR, Article 1.
50 ILO Convention No. 169, Article 6 (a). See also UNDRIP, Article 19.
51 International Law Association, The Hague Conference Report, Rights of Indigenous Peoples (2010), para 51.
52 Human Rights Council, Promotion and protection of the rights of indigenous peoples with respect to their cultural heritage. Study by the Expert Mechanism on the Rights of Indigenous Peoples, 2015, A/HRC/EMRIP/2015/2, para 16.
53 The IPCC was established by the United Nations Environment Programme and the World Meteorological Organization, and subsequently recognized by the UN General Assembly, Protection of Global Climate for Present and Future Generations of Mankind (1988) A/RES/43/53, para 5.
54 Decision 4/CP.20, Report of the Adaptation Committee, FCCC/CP/2014/10/Add.2, at 4.
55 The IPCC, co-convened an international expert meeting on "Indigenous Peoples, Marginalized Populations and Climate Change: Vulnerability, Adaptation and Traditional Knowledge" on 19–21 July 2011 in Mexico City; and sponsored a meeting on "Climate Change Mitigation with Local Communities and Indigenous Peoples: Practices, Lessons Learned and Prospects" on 26–28 March 2012, in Cairns, Australia.
56 The scope of REDD+ includes both activities aimed to reduce forest emissions (such as reduced deforestation, forest degradation and sustainable forest management), as well as activities aimed to increase forest carbon sequestration (such as afforestation and reforestation). See UNFCCC, Decision 1/CP.16, The Cancun Agreements: Outcome of the work of the Ad Hoc Working Group on long-term Cooperative Action under the Convention, FCCC/CP/7/Add.1, para 70.
57 IPBES. 2012. Report of the first session of the IPBES Plenary, IPBES/1/12, p. 29. http://www.ipbes.net/plenary/ipbes-1. Accessed 13 June 2016.
58 IPBES. 2013. Final report and decisions of the second session of the Plenary of IPBES, IPBES/2/17, p. 41. http://www.ipbes.net/plenary/ipbes-2. Accessed 13 June 2016.
59 http://ipcca.info.
60 ALBA is the Spanish acronym for "the Bolivarian Alliance for the Peoples of Our America", an intergovernmental organization formed by 11 Latin American and Caribbean countries.

61 IPBES. 2014. Update on the work of the task force on indigenous and local knowledge systems (deliverable 1 (c)), IPBES/3/INF/2. http://www.ipbes.net/sites/default/files/downloads/IPBES_3_INF_2.pdf. Accessed 13 June 2016.
62 See CBD website, "Participation of Indigenous and Local Communities." www.cbd.int/traditional/general.shtml. Accessed 13 June 2016.
63 Ibid., pp. 26–27.
64 IPBES/3/INF/5.
65 IPBES/3/INF/2, pp. 26–27.
66 IPBES/3/INF/3.
67 United Nations Inter-Agency Support Group (IASG) on Indigenous Issues, "Thematic Paper on the Knowledge of Indigenous Peoples and Policies for Sustainable Development: updates and trends in the Second Decade of the World's Indigenous People." June 2014, paras 5–6.
68 Nagoya Protocol, Article 12.3(a).
69 IPBES/3/L.15.
70 Forest Peoples Programme, "Including Indigenous and Local Knowledge in the IPBES." 2015. www.forestpeoples.org/topics/environmental-governance/news/2015/02/including-indigenous-and-local-knowledge-ipbes. Accessed 27 March 2016.
71 Ibid.
72 See, for instance, CBD COP Decision XII/12(F) in UNEP/CBD/COP/12/29, pp. 91–92.

References

Adger, W.N., Pulhin, J.M., Barnett, J., Dabelko, G.D., Hovelsrud, G.K., Levy, M., Oswald, Spring, U. & Vogel, C.H. 2014. Human Security. In: Field, C.B., Barros, V.R., Mach, K.J., Dokken, D.J., Mastrandrea, M.D., Bilir, T.E., Chatterjee, M., Girma, B., Ebi, K.L., Estrada, Y.O., Genova, R.C., Kissel, E.S., Levy, A.N., MacCracken, S., Mastrandrea, P.R., White, L.L. (eds.), *Climate Change 2014: Impacts, Adaptation, and Vulnerability. Part A: Global and Sectoral Aspects.* Contribution of Working Group II to the Fifth Assessment Report of the Intergovernmental Panel on Climate Change. Cambridge University Press, Cambridge, 755–791.
Berkes, F. 2012. *Sacred Ecology*, 3 edition. Routledge, New York.
Chiarolla, C., Louafi, S. & Schloen, M., 2013. An Analysis of the Relationship between the Nagoya Protocol and Instruments related to Genetic Resources for Food and Agriculture and Farmers' Rights. In: Morgera, E., Buck, M. & Tsioumani, E. (eds.), *The 2010 Nagoya Protocol on Access and Benefit-Sharing in Perspective.* Brill, Leiden, pp. 83–122.
Díaz, S., Demissew, S., Joly, C., Lonsdale, W.M. & Larigauderie, A., 2015. A Rosetta Stone for Nature's Benefits to People. *PLoS Biology*, 13(1), 1–5.
Dutfield, G., 2001. TRIPS-Related Aspects of Traditional Knowledge. *Case Western Reserve Journal of International Law*, 33(2), 233–275.
IPCC, 2007. Climate Change 2007: Mitigation. Contribution of Working Group III to the Fourth Assessment Report of the Intergovernmental Panel on Climate Change. Metz, B., Davidson, O.R., Bosch, P.R., Dave, R. & Meyer, L.A. (eds.) (No. 0521880114 0521705983). Cambridge University Press, Cambridge, UK and New York, NY.
Morgera, E., Tsioumani, E. & Buck, M., 2014. *Unraveling the Nagoya Protocol: A Commentary on the Nagoya Protocol on Access and Benefit-Sharing to the Convention on Biological Diversity.* Martinus Nijhoff, Leiden.

Savaresi, A., 2013. The International Human Rights Law Implications of the Nagoya Protocol. In: Morgera, E., Buck, M. & Tsioumani, E. (eds.), *The 2010 Nagoya Protocol on Access and Benefit-Sharing in Perspective*. Brill, Leiden, pp. 53–81.

Savaresi, A., 2016. Doing the Right Thing with Traditional Knowledge in International Law: Lessons for the Climate Regime. University of Edinburgh Law School. http://papers.ssrn.com/sol3/papers.cfm?abstract_id=2780332.

Soberón, J. & Townsend Peterson, A., 2015. Biodiversity Governance: A Tower of Babel of Scales and Cultures. *PLOS Biology*, 13(3), 1–4.

Sutherland, W.J., Gardner, T.A., Haider, J. & Dicks, L.V., 2014. How can Local and Traditional Knowledge be Effectively Incorporated into International Assessments? *Oryx* 48(1), 1–2.

Tengö, M., Brondizio, E.S., Elmqvist, T., Malmer, P. & Spierenburg, M., 2014. Connecting Diverse Knowledge Systems for Enhanced Ecosystem Governance: The Multiple Evidence Base Approach. *Ambio* 43(5), 579–591.

Wiessner, S., 2013. Culture and the Rights of Indigenous Peoples. In: Vrdoljak, A. (ed.), *The Cultural Dimension of Human Rights*. Oxford University Press, Oxford, pp. 117–156.

11 First thematic assessment on pollination

Between the legitimization of IPBES and tensions regarding the selection of knowledge and experts

Fanny Duperray, Marie Hrabanski and Mohamed Oubenal

Introduction

IPBES, which some experts classify as an "IPCC-like mechanism for biodiversity" (Larigauderie and Mooney, 2010), is a recent institution in search of recognition. One of the keys of that recognition is to be based on IPCC procedures to ensure the credibility[1] of the assessments it wishes to produce. In addition to the credibility guaranteed by the reviewing procedures inherited from the climate research community, such recognition presupposes the nomination of experts that is both fair – in terms of the geographical distribution of the experts – and transparent, so as to guarantee that the results obtained are as legitimate as possible (Mitchell et al., 2006). When IPBES prioritized the main lines of its work program in December 2013, it opted to assess pollination.

The pollination topic offered the advantage of having already been studied in connection with several projects coordinated by the Food and Agriculture Organization of the United Nations (FAO). An appraisal was therefore already available, and some expert networks had been identified beforehand. In addition, the subject had already been touched upon within the Convention on Biological Diversity (CBD), and its coverage in a science-policy interface should make it possible to arbitrate on existing scientific controversies. From that standpoint, IPBES became the ideal arena for coordinating an assessment of the stakes involved in several regime complexes (food security, agriculture, biodiversity, trade, etc.). The long-standing sociotechnical controversy (Bonneuil, Joly & Marris, 2008) regarding the use of neonicotinoids and their effect of causing excess mortality of bees could have been clearly expressed within the recent platform. However, we will show that the institutionalization of controversy though IPBES procedures and the IPBES expert selection process tend to attenuate the controversy on pesticides.

Indeed, in 2014, a controversy was unleashed when the names of the 62 experts selected to produce the first report on "pollination and the role of pollinators" became known. Among them, two were experts working for the agrichemical firms Syngenta and Bayer. Many scientists felt at the time that the selection of these two experts constituted a conflict of interest, which more widely endangered IPBES and its ability to produce a credible assessment. Conversely, some other experts saw this situation as a way of taking different types of knowledge into account. Even others felt that including these stakeholders in the report would also facilitate dissemination of the results within agrichemical companies. This controversy primarily highlighted the close linkages between science and policy in a knowledge selection process like that of IPBES. In fact, the science-policy interface concept is widely used in environmental arenas to qualify hybrid forums where scientists, experts and decision makers interact to select and organize types of knowledge in processes used to frame standards and public action (Koetz et al., 2008; Sarkki et al., 2013). This conceptualization presupposes the existence of two independent "worlds", i.e. science and policy, separated by boundaries. This situation has enabled certain organizations or bodies to play a key role in organizing exchanges between science and policy (Guston, 2001; Van Egmond and Bal, 2011).

The first part of this paper examines this close linkage by analyzing national and transnational policy stakes linked to placing the pollination issue on the IPBES agenda. This enabled us to enhance the way we perceived the role of IPBES in the biodiversity regime complex. In the third section of this chapter, we show how IPBES framed the long-standing sociotechnical controversy on the role of pesticides in the decline of pollinators, notably regarding conflicts of interest that this controversy has given expression to within IPBES. Lastly, we propose to go beyond the science versus policy dichotomy by showing that the process used to select scientific experts was also particularly political insofar as the scientific and policy positions of the experts were taken into account in their recruitment. This selection process leads to attenuation of the debate on the sociotechnical controversy.

Box 11.1 Methodology

In order to examine the selection of experts and controversies regarding the IPBES pollination report, we attended two IPBES Plenary sessions (in Antalya 2013 and Bonn 2015), along with stakeholder meetings during these two events. At these events, we had several opportunities to observe discussions on the selection of the experts and the controversies this process prompted. We also held some more informal discussions during the Bonn 2015 Plenary session to ask the stakeholders present about the controversies. Lastly, we conducted a documentation research on the basis of some official IPBES documents, reports of the Earth Negotiations Bulletin (ENB), comments in the journal Nature and a press analysis via the Factiva database. We also tracked the controversy through an internet search in the blogosphere and via Twitter accounts with the #IPBES3 hashtag during the IPBES Plenary session in Bonn, where there were references to several aspects of the conflict of interest issue. We undertook some research (in the press and on the websites of institutions such as universities, NGOs and governmental bodies) on the experts involved in Chapter 2, as it focuses more on the causes of the decline in pollinators. We also interviewed five of them and searched the Web of Science for articles by the Chapter 2 experts linked to the pesticide issue.

Placing pollination on the IPBES agenda – a strategic choice to legitimize the IPBES coordination role in the biodiversity regime complex

Pollination has been chosen by IPBES as the number one topic for a thematic assessment,[2] which will be completed and made public in 2016, during the Platform's 4th Plenary session. Over and above considerations regarding the visibility of the subject, we show that pollination helps consolidate the recognition of IPBES as a forum for coordination between several regime complexes while strengthening its links with CBD.

Pollination – a visible stake that responds to national policy challenges

Pollination was thus pinpointed as a priority by the Multidisciplinary Expert Panel (MEP) and the IPBES Bureau. IPBES is therefore striving to fulfill the requirements of international agreements on biodiversity, notably CBD, thereby making pollination a "flagship" (IPBES/2/INF/9) to demonstrate the relevance of IPBES. When the work program was adopted at Antalya in December 2013, the MEP/Bureau felt that pollination should be the top priority, whatever the budget available to IPBES.

The visibility objective was presented by the Bureau and the MEP as an argument for the choice of the theme, as specified in a report on "receiving and prioritizing requests, contributions and suggestions put to the Platform" (IPBES /2/3). This report stated that the MEP and the Bureau

> considered that high priority should be given to a small number of thematic assessments to be undertaken during the period 2014–2016, because this will provide the opportunity to quickly demonstrate the Platform's utility to scientists, policymakers, decision makers and other stakeholders. Several submissions and comments on the work program highlighted the need for early, high-visibility products from the Platform. These thematic assessments will also provide a testing ground for mobilizing the scientific community and holders of indigenous and local knowledge; developing capacity and creating policy support tools; and providing

valuable insight into processes and procedures for the more complex regional and global assessment processes.

(IPBES/2/3)

The choice of topics on the assessment agenda was based on requests, prioritized by MEP and the Bureau. Of the 52 requests received, four concerned pollination (IPBES/2/INF/9), i.e. the ones from New Zealand, Norway, CBD and the Convention on Migratory Species (CMS). Pollination was therefore not a subject in strong demand from the states and stakeholders. Other subjects received a larger number of requests, such as the sustainable use of biodiversity, but which entailed issues that are more difficult to circumscribe. Despite this, the choice of pollination also seems to correspond to media concerns and public opinion in industrialized countries, notably in Europe, Canada and USA, as confirmed by the interviews carried out. Furthermore, EU policy makers declared a moratorium from 2013 to 2015 on the usage of neonicotinoids in pollinator-dependent crops.[3] In doing so, they have taken uncertain knowledge and suggestive evidence of harm from laboratory and field studies seriously – a false-positive policy orientation that indicates a preference for bearing the costs of being wrong about the harm posed by these chemicals rather than overlooking that harm. In USA, there has been considerable movement on this issue at the US Environmental Protection Agency (EPA), including new labelling requirements and pollinator risk assessment guidelines, as well as likely restrictions on new outdoor uses of these chemicals in the absence of additional effects-data on developing honey bees, even though EPA has to date refused to take action to prohibit or restrict current usage patterns regarding these chemicals in general, and in specific usage contexts (Suryanarayanan and Kleinman, 2013).

Authorities in developing countries, apart from a few, feel much less concerned by the issue, since the use of inputs there is much more limited in most cases than in industrialized countries. In addition, public opinion in developing countries seems much less sensitive to the health, economic and social problems that arise as a result of the decline in pollinators. IPBES is undoubtedly influenced with regional and national dynamics that guide the themes of the Platform work program, even though the states had not explicitly called for pollination to become the subject of the first report.

Pollination – a stake for recognition of the coordination role of IPBES in the biodiversity regime complex

Over and above the high visibility of the pollination theme, it can also be seen that IPBES opted for a subject that meets CBD requirements and has thereby sought to strengthen the link between IPBES and CBD, along the lines of IPCC/UNFCCC relations. Indeed, pollination was first added to the CBD agenda at the end of the 1990s, at the time of the Sao Paolo Declaration on Pollinators in 1998, which recommended that CBD create an International

Pollinator Initiative. However, the issue appeared too thorny to be dealt with in an arena that was "too" political. The CBD strategy was therefore to send this thorny issue to the scientific arenas first. To that end, CBD dealt with the issue by launching the International Pollinator Initiative (IPI) that was implemented on regional levels. IPI was adopted by CBD[4] during COP5 in 2000 as the developing countries felt that this international convention should focus on food security issues in order to be seen as legitimate in the eyes of developing countries. CBD entrusted FAO with coordinating the process and submitting a progress report to each COP. An Action Plan for IPI was adopted for COP6 in 2002, which intended to guide some regional initiatives which were launched after the Sao Paulo declaration. The Brazilian Pollinator Initiative (BPI), the African Pollinator Initiative (API), the North American Pollinator Protection Campaign (NAPPC) and the European Pollinator Initiative (EPI) joined IPI when it was officially launched at the international level by CBD.

The selection of experts in the international scientific programs was closely linked with that of IPBES, with a large share of the people involved in the IPBES report on pollination thus being part of the International Pollinator Initiative network. Likewise, many scientists involved in the European projects ALARM, STEP, LIBERATION, and now SUPER B, are also experts who were selected for the IPBES report.[5]

Pollination is also an issue at the interface between the biodiversity regime complex, agriculture and trade. Indeed, the decline in pollination worldwide is a cause for concern not only for safeguarding biodiversity, but also regarding the impact that it might have on agricultural production in multiple countries and on an international scale, as illustrated by the fact that the pollination issue has been placed on the agenda within international organizations and conventions (CBD, FAO). This shows how IPBES could coordinate the production and selection of knowledge in several regime complexes. Strategically, it was also a way for IPBES to benefit from the appraisal already undertaken by FAO, while asserting its primacy over this issue, so that it would not be viewed solely in terms of agricultural stakes, within the agriculture regime complex, but also via deliberations in terms of stakes linked to biodiversity. Lastly, in this way IPBES also strengthened its links with CBD insofar as pollination is completely in line with the Convention requirements.

In terms of economic and commercial stakes, the pollination assessment led to a large-scale summary of existing results relative to the economic assessment of the ecosystem service represented by pollination. Moreover, it follows on from earlier reports, such as those concerning the Millennium Ecosystem Assessment (2005)[6] and The Economics of Ecosystems and Biodiversity[7] (TEEB, 2008). MA thus devoted a sub-section to pollination, while TEEB presented a case study of the economic value of beekeeping in Switzerland, which took another look at the results of a study by Gallai and colleagues which revealed that the overall economic value of pollination by insects amounts to US $153 billion per year (Gallai et al., 2009).

IPBES – a new arena for expressing sociotechnical controversy on the role of pesticides in the decline of bees?

When the recent IPBES Platform decided in 2013 to produce a report on pollination and the role of pollinators, it could have become a new arena for expressing the old sociotechnical controversy on the role of pesticides in the decline of pollinators. Beekeepers, firms, federations and civil society end up defending opposite interests. Some virulent viewpoints have been expressed and numerous tensions have emerged since the 1990s, notably between certain beekeeping federations and agrichemical companies producing pesticides. The latter feel that proper use of neonicotinoids does not have any impact on bee mortality. Some more or less restrictive measures have been taken by national policy makers to try and reconcile economic issues and ecological stakes (moratoria, incentive measures, bans, etc.) However, within IPBES, the controversy on the role of pesticides in causing excess bee mortality was expressed through the process used to select experts of the report, which finally shifted the focus onto IPBES procedures rather than an in-depth debate on the controversy. Following a selection process, 62 experts (IPBES/3/INF/5) were assigned to draft a report on pollination and the role of pollinators. Of them, two were experts working for agrichemical firms, i.e. Syngenta and Bayer. Some people felt that their presence might introduce bias into the report, while others saw in it a way of representing different types of knowledge in the report, and some felt that by including these stakeholders in the report, it would facilitate dissemination of the results, including within agrichemical firms. After analyzing the timeline of the controversy surrounding the nomination of the experts for the IPBES pollination report, we show how pollination fashioned the deliberations regarding the procedures used to manage conflicts of interest amongst IPBES experts.

Timeline of the controversy on the involvement of experts from agrichemical firms in the IPBES pollination report

The timeline of events marking the selection of the IPBES pollination experts combines events emerging in both the scientific and media arenas (press, blogs and social networks).

It was in June 2014 that the first list of experts selected to draft the pollination report was made public. A second event, although not directly linked to IPBES, took place in June 2014 when a group of 29 scientists from the Task Force on Systemic Pesticides[8] published a report entitled 'Worldwide Integrated Assessment on Systemic Pesticides' (WIA) which was released in the form of several scientific articles published in an issue of the peer-reviewed journal *Environmental Science and Pollution Research* in 2015 (van Lexmond et al., 2015). Based on over 1,000 publications on neonicotinoid pesticides, WIA concluded that there can be no doubt that neonicotinoids are a threat to biodiversity and bees and they should be eliminated as soon as possible.

Of the 29 authors of the WIA report, some were also among the experts nominated, but not selected by IPBES, to take part in the pollination assessment.

Then, in December 2014, an article co-signed by three authors was published in *Nature* denouncing the presence of two representatives from agrichemical firms on the list of experts selected for the IPBES pollination report[9] (Hochkirch, McGowan & van der Sluijs, 2014). The three authors criticized what they felt to be a conflict of interest, in that the two experts from the private sector could not make a credible assessment as they were paid by the firms. The authors of that article called for "more explicit rules for the selection or nomination of experts in IPBES" and felt that the experts from the industry should not be among the co-lead authors (CLAs) or lead authors (LAs) of the report. One of the three authors was a member of the Task Force on Systemic Pesticides scientific network.

It was also at that time that some scientists used their blog and/or tweets to express their concerns about the nominations of the two experts from the private sector[10] or to support the inclusive stance of IPBES.

A month later, in January 2015, A. Larigauderie – the IPBES Executive Secretary[11] – and colleagues responded in *Nature* to the previous article claiming that the rules were sufficiently explicit and transparent as the experts were nominated by governments and stakeholders and selected by the MEP members (Larigauderie, 2015). The authors also used the credibility and legitimacy of IPCC to issue a reminder that some private sector representatives were also part of IPCC and the International Assessment of Agricultural Knowledge, Science and Technology for Development (IAASTD), and that multi-stakeholder assessments are thus fully legitimate. Lastly, the authors further pointed out that the issue of declaring conflicts of interest was to be shortly debated at the Plenary session in Bonn due to be held in January 2015.

The debates emerging in the scientific arenas, notably in *Nature*, rapidly spread and also emerged in some more media-related arenas. A degree of porosity exists between these two arenas, but it is a matter here of highlighting the fact that the controversy regarding the choice of experts had burst out of the strictly scientific framework and into the media. An article published in the highly regarded French newspaper *Le Monde*, in November 2014 by Stéphane Foucart, also denounced this situation of interests, even before the declaration made in *Nature*. More precisely, Foucart analyzed the expert selection process, highlighting the presence of scientific experts from the private sector, whose scientific results were more than open to criticism.[12] The journalist concluded from this that IPBES had sent the wrong signal by selecting this type of representative. On the other hand, a blogger felt that those who were worried about the involvement of the two representatives from the private sector in the report, and their possible conflicts of interest, were not worried about the involvement of scientific activists linked to environmental NGOs or anti-pesticide NGOs in other assessment agencies.[13]

The controversy on the selection of experts was to be debated within IPBES and fueled a more general debate in which the issue was viewed from the conflict of interest angle – which ultimately amounted to IPBES controlling and retaking the upper hand in the communications surrounding the pollination report.

Institutionalizing the expert nomination controversy – negotiating the rules on conflicts of interest

The IPBES-3 Plenary session opened in Bonn in January 2015 in a context of declarations in the scientific and media arenas. Institutionalization of the controversy by the IPBES platform and its appropriation by national delegations within IPBES contributed to shifting the problem towards issues of legal procedures for managing conflicts of interest, thereby making it less conflictual. This does not mean that disagreements between the delegations did not occur, but they were channeled by the format of the debates in the Plenary session and the Platform mode of governance. Thus, despite the bipolarization of positions regarding conflicts of interest, the debates in the Plenary session reached a consensus on the way to manage conflicts of interest.

In preparation for the discussions during the Plenary session on conflicts of interest, several documents were proposed by the Secretariat based, down to the final detail, on IPCC procedures (Edenhofer, 2011). However, many delegations identified other options. Several informal and/or regional meetings were held in succession during the week of the Plenary meetings in order to pinpoint the positions and stakes regarding the management of conflicts of interest. A distinction could be made between the different options proposed by the delegations by positioning them along a continuum with, on one side, a pole borne by the European Union, notably France, Belgium and Norway, who wished to adopt rules that were as transparent as possible which, according to those delegations, would make it possible to avoid media controversies. Those delegations also wanted it to be possible for an organization or individual person to report a conflict of interest to the IPBES committee responsible for examining the issue. For this first pole, it was thus a matter of going further than the procedures proposed by IPCC as according to some interviews of delegation representatives. These also wished to maintain an IPCC rule imposing the publication of conflicts of interest regarding experts who would retain their functions due to their "irreplaceable contribution" to the Platform (Shapiro et al., 2010).

The other pole was represented by Switzerland and the USA. This second pole took a much more flexible view regarding conflicts of interest. These delegations wished to protect the privacy of experts. Unlike the delegations close to the first pole, those of the second pole felt that the fact of being able to report a conflict of interest to IPBES would create a climate of suspicion.

Interestingly, for the pole calling for transparency, being as transparent as possible was a token of IPBES legitimacy and would make it possible to reassure scientists and civil society in the face of potential controversies emerging in the scientific and media arenas regarding the selection of the experts. On the other hand, those who feared that a climate of suspicion would be created felt that IPBES had, on the contrary, managed very well to deal with the problem and the conflict of interest accusations concerning pollination, hence there was no need to strengthen the conflict of interest management procedure.

These debates led to a solution that was unprecedented in comparison with the IPCC procedures – the creation of a Conflict of Interest Committee, comprising three members of the Bureau, including a vice-chairman, five members nominated by the members (one per region) and a legal expert from the United Nations Environment Programme. Anyone could submit a duly argued request to examine a potential conflict of interest to the Bureau, which in turn would refer the case to the Committee for appraisal.[14] This procedure did not go as far as the position taken by those who wanted any member or observer of the Platform to be able to refer conflicts of interest to the Committee, or wanted these new rules on conflicts of interest to be retroactive, which in this case would have made them applicable to the experts from Syngenta and Bayer.

In this way the sociotechnical controversy on the role of pollinators finally mainly focused on IPBES procedures. IPBES did not really become a new forum to express it because the first IPBES goal seems to be as legitimate and established as IPCC. This trend shifted the debate to be less focused on strictly scientifically and political issues concerned by the sociotechnical controversy.

Process for selecting scientific experts on pollination – challenges and avoidance of debate on the controversy?

An analysis of the formal and informal dimensions of the expert selection process showed that the procedures planned by IPBES left plenty of room for manoeuver for some stakeholders, notably the report chairs and CLAs, to adapt or even get round the official selection rules in order to include experts whom they felt it was important to work with. In addition, the formal rules fostered the idea of a geographical balance by major regions and a gender balance, but these formal recommendations were also largely bent to meet others' needs.

Respecting the selection procedure rules while circumventing them – a challenge for selecting experts

Officially, MEP played the most decisive role in the expert selection process. The documents stipulated that the "report co-chairs, coordinating lead authors, lead authors and review editors are selected by the Multidisciplinary Expert Panel from the lists of nominations" (IPBES/2/17, 3.6.1, see also the section on IPBES operations, Chapter 3). MEP, advised by members of the Bureau (IPBES/3/INF/5), thus first selected certain report co-chairs and coordinating lead authors who took part in selecting some other members of the group of experts on pollination. However, in practice, the CLAs and report co-chairs played a greater role than suggested in the texts. By leaving more room for manoeuver for the CLAs and report chairs, the process made it possible to respect the formal selection rules while providing IPBES with access to the collaboration networks of those experts.

This was notably the case when one of the report co-chairs suggested to some experts that they enter their candidature via their country to join the team that was going to be formed. Likewise, two other interviewed experts specified that CLA were first selected and then took part in selecting the lead authors of their chapters. Knowledge of the selection rules also made it possible to avoid that some people qualified for drafting the document were turned away due to over-formal principles. One of the female experts working as a researcher in a country that is not an IPBES member thus entered her candidature via a member country in order to take part in the pollination assessment. This way of doing things was proposed to her by a report co-chair. This technique would therefore seem to make it possible to get round the different problems that might arise in subsequent assessments if renowned researchers are not nominated by their respective countries.

Adapting the procedures of geographical diversity, gender and knowledge systems

While the report chairs and CLAs were able to bring together some researchers skilled in scientific assessments of pollination, they also had to take care that geographical diversity, gender and knowledge system principles were respected.

Some expert selection principles are set out in the procedures. This concerns the excellence of the candidates and the relevance of their expertise. This criterion is combined with a requirement for gender balance, diversity of knowledge systems, geographical representation, with appropriate representation of experts from developing and developed countries and countries with economies in transition. However, it was difficult in practice to ensure a balance of these three parameters. Regarding the gender distribution balance, for instance, it can be seen that 32% of the experts in the group were women (IPBES/3/INF/5). Although IPBES representatives vaunt the size of that figure compared to that of IPCC, i.e. at around 20% (IPCC/IPBES meeting in Paris), it is far from parity. In addition, our interviews revealed the centrality of certain male experts from developed countries, since some female researchers considered that it was in itself an honor to work with these renowned scientists. If we consider the balance in the geographical distribution according to UN regions, it was heavily weighted in favor of regions including Western Europe, North America and Australia, which alone accounted for 39% of the selected pollination experts. On the other hand, the UN Eastern Europe region was less represented by 3% of the experts, followed by Africa with 14%, Asia-Pacific with 16%, and then Latin America and the Caribbean with 27% (IPBES/3/INF/5).

Lastly, in terms of the diversity of knowledge systems, it should be pointed out that two levels exist – first taking indigenous and traditional knowledge into account, and then integration of social sciences and not only natural sciences. On this second point, emphasis was placed on the economics discipline as several researchers were selected to take part in drafting chapter 4 devoted to economic methodologies for the pollination assessment. As regards chapter 5, devoted to the non-economic assessment, it was to include indigenous and local knowledge as well as inputs from social science research. When taking indigenous and traditional knowledge into account, even though a parallel process was set in place, there were a few hesitations concerning their inclusion in the report. At one meeting of experts, some scientists wanted to restrict these different types of knowledge to chapter 5, devoted to the non-economic assessment, but in the end it was those who wanted to incorporate them in all phases of the process who ultimately succeeded in getting their way. However, the conducted interviews highlighted that the experts responsible for clarifying indigenous knowledge in the different chapters had difficulties in making their voices heard to the same degree as experts producing so-called scientific knowledge. Their knowledge seemed to be taken into account when so-called scientific knowledge was lacking. Traditional knowledge was also more favorably welcomed by scientific experts from developing countries than by those from industrialized countries. Generally within IPBES, as in its pollination group of experts, the method by which these different types of knowledge would be concretely incorporated was a subject of debate.

Our analysis will now focus on the experts selected for chapter 2, 'Drivers of Change of Pollinators'. One of the purposes of this chapter was to determine the effects of different risks on pollinators and pollination. As mentioned in the initial presentation document of the report, "it will assess direct drivers of change in pollination, including the risk posed by climate change, invasive species and diseases, land-use changes, changing agricultural practices, and the use of chemicals including fungicides and insecticides" (IPBES/3/INF/5). We therefore decided to analyze this chapter because it was the one at the center of the controversy on the expert selection process presented in section 3. In fact, the subjects covered by the other chapters (pollinator diversity, assessment of the decline in pollinators, assessment of the value of pollination) were not the focus of scientific debates that gave rise to controversies.

Scientific and public declarations of experts taken into account in the expert recruitment process – focus on chapter 2

Chapter 2 was the most discussed chapter of the pollination report. In this respect, the recruitment of experts was a particularly sensitive subject. In USA, Suryanarayanan and Kleinman showed that "a certain set of research norms and practices from agricultural entomology came to dominate the investigation of the links between pesticides and honey bee health, and how the epistemological dominance of these norms and practices served to marginalize the knowledge claims and policy positions of commercial beekeepers in the Colony Collapse Disorder (CCD) controversy" (Suryanarayanan and Kleinman, 2013). Among the 12 experts involved in drafting this chapter of the report, it can already be said that none of the experts selected were specialized in the social sciences. Likewise, no representatives or experts directly represented the knowledge of indigenous peoples. Of the 62 experts of the report on pollination overall, there were fewer than five economists and only one anthropologist. Despite the initiatives of the Secretariat, indigenous and local knowledge (ILK) played little part in the pollination report. Over and above the role of ILK, the expert recruitment process also revealed the stakes surrounding the selection of relevant knowledge for investigating a controversy. We propose to classify the authors of chapter 2 – devoted to defining the risks of the decline in pollination – in three categories according to their proximity to the controversy on the involvement of insecticides in bee mortality (see Figure 11.1). An initial group of players were at the heart of the controversy, since it contained some individuals who made public and clearcut pronouncements in the media and carried out scientific research on the subject. A second group of players took a stance of caution and nuanced positions when speaking in the media, but also carried out research on the issue of the role of insecticides in bee mortality. A third group consisted of researchers, some of whom worked on the other causes of bee mortality (climate, landscape structuring, etc.) but were not working on insecticides.

Clearcut position on controversy

The first group of researchers at the heart of the controversy on the implication or not of insecticides in bee mortality consisted of two experts. Both of them

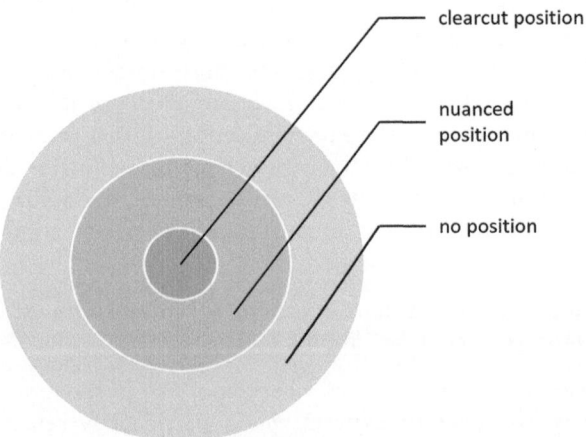

Figure 11.1 Chapter 2 authors' positions

made clearcut and public declarations in the controversy but, while one declared that insecticides are not linked to bee mortality, the other considered, on the contrary, that insecticides are the main factor in the bee mortality.

Another aspect distinguished the two experts from each other – the first was more specialized on the insecticide issue than the second. Thus, although they were both at the heart of the controversy, the first had published 22 articles on the question of insecticides while the second had only published six. Although he had studied pesticides during a training course at the beginning of his career, he then switched the focus of his studies and began working on economic development and small-scale honey producers.

The clearcut position that tended to limit the effects of insecticides on the decline in bees was therefore represented by a researcher who was more specialized, in terms of publications, than the researcher taking the other position. Yet there are still some experts who consider that insecticides are one of the main causes of the decline in bees, while being specialists on this issue (e.g. see the Task Force on Systemic Pesticides network). However, they were not involved in chapter 2 on the 'Drivers of Change of Pollinators' in the IPBES pollination assessment.

A third aspect distinguished these experts from each other. The work of the expert who felt that pesticides were not dangerous had been challenged by scientists, and this point is regularly highlighted by some NGOs and the media. Moreover, the latter had been suspected on several occasions of conflicts of interest by various NGOs and some journalists, who accused him of producing biased data, ultimately benefiting pesticide firms. The other expert was not accused of producing biased data by the stakeholders in the debate.

Some nuanced positions

The second group of players was further away from the controversy than the first group, but was nonetheless implicated. It included four researchers who had dealt with the question of pesticides via some of their publications, and some of their appearances in the media. However, unlike the previous group, this group took a cautious stance on the pesticide issue.

These researchers had published on pesticide questions; two of them had taken part in some assessments on the neonicotinoid issue, while the other two had produced research on the pesticide question. However, none of them were specialized in this field. All of these researchers had broader expertise, in terms of the number of publications, on other factors involved in the decline of pollinators, such as landscape structuring, the climate, and bee pathologies. As regards speaking to the media, these researchers had expressed themselves in the newspapers more or less frequently, including in media with a relatively wide distribution. For example, one of the experts, at the beginning of his career, had only spoken to some media publishing scientific topics for a general audience. The others had made declarations in media with a wider audience. In these public declarations, the expert scientists did not assert that pesticides played a definite and major role in the decline of bees. Some of the experts highlighted the scientific uncertainties persisting on the subject, and emphasized the multi-factor nature of the causes, whilst acknowledging that the results of the studies on pesticides were worrisome. Some of them did not speak out on the subject directly, but spoke about assessment methods, and research that needed to be stepped up. This cautious attitude on pesticides, as seen in the analysis of the previous group, was not the stance of all researchers. One might wonder to what extent this choice not to make clearcut declarations on the role of pesticides in the decline of bees corresponded to a strategy adopted by these experts in response to institutional or political constraints. For example, we found different declarations by one of these researchers depending on the audience he was addressing. In certain media, he asserted that pesticides very probably have an impact, in interaction with other factors, while in other media, he spoke about concerns about pesticides. Then, before a political body where he was a guest expert, he stated that varroa mites (a parasite affecting domestic bees) were, without doubt, the main factor in the decline of bees, and that he had no opinion on pesticides. Another of the experts signed a document intended for politicians estimating that neonicotinoids have an impact and their use should be reduced as quickly as possible. However, on an individual basis, he had never defended that assessment and its recommendations in the media, unlike the other authors. Yet, this expert had taken policy positions on other subjects, such as the use of GMOs in rice plantations, or light pollution. This stance seemed to be justified by a will to appear neutral, as confirmed by the interview conducted, insofar as the researchers who take a position are regularly accused by pro-pesticide stakeholders of conflicts of interest. The fact that these experts highlighted the uncertainties existing in this field may therefore be linked to the scientific debate, but also to a desire not to take a position in the controversy.

No side taking in the controversy

Four experts did not make any public declarations on the controversy concerning the decline in bees. A major distinction could be seen among these four experts, since three of them had expertise in bee biology, while the last had no such expertise. That expert was specialized on the question of climate impacts on agriculture, development and water management issues.

Of the three experts working on bee biology, the first had few publications and worked on geophylogeny, which is not directly linked to the decline of bees. This author, like the previous one, was quite peripheral to the controversy. The other two worked specifically on the causes of the decline (bee pathogens and the role of landscapes) but not on the role of pesticides in that decline. These two experts had published abundantly on their research subjects. However, none of these three experts had made any public declarations on the causes of the decline in bees, and consequently on the role of pesticides.

The analysis of publications and public declarations of the experts showed that out of the 10 experts in chapter 2, the most debated chapter, a small proportion of the experts had made declarations on the controversy – considering that pesticides did little harm to bees or, on the contrary, that they were dangerous. However, the expert selection process did not lead to any type of confrontation between scientists stating with certainty the effects of pesticides, and those scientists demonstrating an absence of their effect. In chapter 2, no scientists with the same degree of specialization on the issue came to opposite conclusions. Of the two experts who supported clearcut positions, one was less specialized than the other on the insecticide question. The second group had the highest number of experts who held nuanced and cautious positions on the causes of the decline. Lastly, the third group was composed of heterogeneous experts, one had no expertise on the issue, the other had worked on pollination, but not on the role of pesticides in the decline of bees, and none of them had made public declarations on pollinators. These researchers were therefore more distant from the controversy.

In mixing these three different types of experts, the IPBES expertise process is based on the legitimacy of the specialized experts who have a clearcut position, of the specialized experts who have a nuanced position and of the experts with a more distanced position. In this way, this process enables debate on the controversy, but the strongest positions are attenuated because of the experts with a more distanced position. There were no camps that put forward some serious arguments and clashed with any other camp. In this way, the expertise process contributes to attenuate the controversy and the expression of it and discussions on it.

Conclusion

Based on the first elements available on the pollination assessment produced by IPBES, this chapter has highlighted the stakes involved in the production of expertise in the new political science interface on biodiversity. We have first shown that the choice of pollination proved to be particularly strategic for the Platform, in so far as this subject lies at the interface between two regime complexes (biodiversity and agriculture). IPBES chose to focus on pollination in coordination with international conventions, notably CDB, while also making use of FAO work on the issue. The Platform thus strengthened its links with CBD, whilst positioning itself as a stakeholder capable of coordinating an initiative on scientific stakes involved in two regime complexes.

Lastly, the chapter also showed how IPBES missed or avoided the opportunity to become a really new arena for showcasing the sociotechnical controversy on the role of pollinators by focusing on procedures. We showed that the controversy soon becomes focused on the issue of conflicts of interest. The IPBES selection process thus did not lead to an in-depth debate on the role of pesticides, i.e. no experts have directly expressed their position. In this way, the institutionalization of controversy though IPBES procedures and processes contributes to attenuating the controversy on pesticides. IPBES seems to have managed to overcome criticism directed at it notably by taking advantage of the notoriety of IPCC, and by the procedures it has developed in this field. The necessity to ensure the legitimacy of IPBES could result in avoidance of the sociotechnical controversies.

Overall, the selection of scientific knowledge and experts has been largely permeable to the political stakes. Moreover, the place granted to the knowledge-holders, whether they be from the private sector or indigenous peoples, is far from being settled, as illustrated by the difficulties in integrating these different types of knowledge.

Notes

1 We refer, in this article, to the concept of credibility defined by Mitchell et al. (2006: 317) such as: "convincing actors that the facts, theories, ideas, models, causal beliefs, and options contained in an assessment are 'true', or at least a better guide to how the world works than competing information".

2 At the outset, IPBES called this type of assessment a "fast-track thematic assessment". That designation has since been dropped.

3 Neonicotinoids are among the most widely used pesticides in agriculture. They are systemic, hence present throughout the plant, and neurotoxic. According to different scientific publications (van Lexmond et al., 2015), they are consequently considered to be dangerous for pollinators. Many beekeeping associations and environmental NGOs have been very active in condemning their use.

4 The regional initiatives were launched after Sao Paolo, and were then attached to the IPI network once it was formally adopted by CDB.

5 Thus, the two co-chairs of the IPBES report were from these scientific networks, and 15 of the 17 coordinating lead authors are also part of those networks. UN organizations such as FAO were involved in the selection process. Two FAO pollination experts were also involved in drafting the IPBES report.

6 www.unep.org/maweb/documents/document.295.aspx.pdf

7 www.teebweb.org/our-publications/teeb-study-reports/ecological-and-econom ic-foundations/#.Ujr1xH9mOG8

8 The Task Force on Systematic Pesticides was set up in 2012, following the Appeal of Notre Dame de Londres in 2009, during which a group of entomologists and ornithologists met to express their worries about the steady decline of bees. The task force is currently linked to two International Union for the Conservation of Nature commissions, the Species Survival Commission (SSC) and the IUCN Commission on Ecosystem Management (CEM).

9 www.nature.com/nature/journal/v516/n7530/full/516170c.html

10 See for example one of the tweets by Vander Sluijs: https://twitter.com/Jeroen_vdSluijs/status/543306274637885441.

11 On behalf of herself, the two co-chairs and the pollination assessment coordinator.

12 www.lemonde.fr/biodiversite/article/2014/11/10/les-experts-le-contrat-de-defiance_4521395_1652692.html

13 http://risk-monger.blogactiv.eu/2014/12/16/the-bee-sting-activist-scientists-and-the-a buse-of-power/#.VQlYiOHpW8Y
14 www.fondationbiodiversite.fr/images/stories/telechargement/CR_IPBES3_final.pdf

References

Bonneuil, C., Joly, P.-B. & Marris, C. 2008. Disentrenching experiment the construction of GM – crop field trials as a social problem. *Science, Technology & Human Values*, 33, 201–229.

Edenhofer, O. 2011. Different views ensure IPCC balance. *Nature Climate Change*, 1, 229–230.

Gallai, N., Salles, J.-M., Settele, J. & Vaissière, B. E. 2009. Economic valuation of the vulnerability of world agriculture confronted with pollinator decline. *Ecological Economics*, 68, 810–821.

Guston, D. H. 2001. Boundary organizations in environmental policy and science: an introduction. *Science, Technology, and Human Values*, 399–408.

Hochkirch, A., Mcgowan, P. J. K. & Van der Sluijs, J. 2014. Publishing: Biodiversity reports need author rules. *Nature*, 516, 170–170.

Koetz, T., Bridgewater, P., Van den Hove, S. & Siebenhüner, B. 2008. The role of the Subsidiary Body on Scientific, Technical and Technological advice to the Convention on Biological Diversity as Science Policy interface. *Environmental Science and Policy*, 505–516.

Larigauderie, A. 2015. Pollinator assessment: IPBES responds on conflicts of interest. *Nature*, 517, 271–271.

Larigauderie, A. & Mooney, G. 2010. The Intergovernmental science-policy Platform on Biodiversity and Ecosystem Services: moving a step closer to an IPCC-like mechanism for biodiversity. *Current Opinion in Environmental Sustainability*, 2, 9–14.

Millennium Ecosystem Assessment 2005. *Ecosystems and Human Well-being: Synthesis.* Washington, DC, USA: Island Press.

Mitchell, R. B., Clark, W. C., Cash, D. W. & Dickson, N. M. 2006. *Global Environmental Assessments: Information and Influence.* Cambridge: MIT Press.

Sarkki, S., Niemelä, J., Tinch, R., Van den Hove, S., Watt, A. & Young, J. 2013. Balancing credibility, relevance and legitimacy: a critical assessment of trade-offs in science–policy interfaces. *Science and Public Policy*, sct046.

Shapiro, H. T., Diab, R., De Brito Cruz, C., Cropper, M., Fang, J., Fresco, L., Manabe, S., Mehta, G., Molina, M. & Williams, P. 2010. Climate change assessments: Review of the processes and procedures of the IPCC. *InterAcademy Council, Amsterdam.*

Suryanarayanan, S. & Kleinman, D. L. 2013. Be(e)coming experts: The controversy over insecticides in the honey bee colony collapse disorder. *Social Studies of Science*, 43, 215–240.

TEEB 2008. *The Economics of Ecosystems and Biodiversity* – Interim Report. European Communities.

Van Egmond, S. & Bal, R. 2011. Boundary configurations in science policy: modeling practices in health care. *Science, Technology & Human Values*, 36, 108–130.

Van Lexmond, M. B., Bonmatin, J.-M., Goulson, D. & Noome, D. A. 2015. Worldwide integrated assessment on systemic pesticides. *Environmental Science and Pollution Research*, 22, 1–4.

12 Conclusion

The emergence of complex scientific governance

Philippe Le Prestre

Introduction

Created under the combined pressure of several middle powers and various groups of stakeholders, and despite strong initial skepticism from several major states, IGOs, and even NGOs, the Intergovernmental Platform on Biodiversity and Ecosystem Services (IPBES) represents a remarkable example of the contemporary dynamics of international institution-building. Its basic features evolved during the negotiation process: the linear model of science-policy relations became dialectical, the range of relevant knowledge expanded, as did the diversity of stakeholders. Thus, it also constitutes an experiment in new forms of governance, including attempts to invent ways of approaching and conducting assessments and a new science-policy dialogue.

IPBES strives to reconcile three approaches to biodiversity protection: scientific (biology and ecology, with a focus on conservation and knowledge building), utilitarian (ecological economics, with a focus on ecosystem services and the science-policy interface) and cultural (anthropology and politics, with a focus on institutions and on the protection of knowledge and political rights). Although the advent of IPBES represents a political alliance of sorts, especially between the first two, it also falls within the larger debate on what should form the basis of decisions regarding biodiversity and, more generally, on the proper relationship between societies and nature. This debate, which is reflected in the triple narrative identified by Alice Vadrot (see chapter 3), has affected and will continue affecting the definition of IPBES procedures, starting with the development of a "coordinative discourse" designed to guide decisions and act as a standardization device (see chapter 7).

Three basic assumptions that have driven the genesis of IPBES, shaped its mission, determined its structure, and guided its operations correspond to these approaches. These assumptions have not necessarily been widely shared – which explains the time it took for IPBES to emerge – and they still face contradictory interpretations regarding their importance and implications. They pertain to filling knowledge gaps, improving the science-policy interface, and including all stakeholders in the development of policy-relevant knowledge. A brief examination of them sheds some light on the two questions posed at the beginning of this book: (1) What explains the emergence of IPBES in the context of a global biodiversity issue-area that is both institu-tionally fragmented and complex? (2) How does IPBES manage tensions between the openness ambition, the knowledge selection process and the production of global assessments by experts, i.e. tensions arising from directions taken in response to those assumptions?

Assumption #1: there is not enough science

This assumption reflects the view of natural scientists that biodiversity is eroding faster than the increase in our knowledge of it, thus impeding societies' ability to make informed choices. As detailed in the introductory chapter, this conjecture underlies the main objective of IPBES, whose role is to identify research gaps, produce syntheses, and build scientific consensus, while not generating new knowledge *per se*.

This assumption has two dimensions. First is the belief that there is not only a lack of basic knowledge but there are also gaps in our capacity to put together and make collective sense of existing knowledge (see chapter 3). IPBES largely approaches the biodiversity issue as being related to deficient or unavailable science, not to values and behavior. It assumes that knowl-edge will impact values, which in turn will shape behavior. Holders of this assumption are not primarily concerned with the impact of institutional fragmentation on policy *per se*, but rather with building a scientific consensus and expanding knowledge. Yet, the further assumption of the universal character of science leads to the belief that science can provide the basis for coordinated action. The objective, therefore, becomes the development of a multidisciplinary "biodiversity science" informed by other types of knowledge. The second associated dimension is the unequal weight given to various sources of knowledge, from the social sciences and humanities to those deemed non-scientific, such as local and indigenous knowledge.

Assessments serve to mobilize the scientific community, states and various stakeholders, with the aim of providing a common framework for the development of new knowledge and actions. The importance and shape of their construction are rooted in the successes and shortcomings of the Intergovernmental Panel on Climate Change (IPCC), a question that is speci-fically addressed in chapters 3, 5 and 6. Whether as model or counterpoint, IPCC has been a reference point for IPBES from its inception. As detailed in chapter 4, assessments represent the main activity of IPBES, marshalling about 44.5% of its estimated budget, with 31% devoted to global as opposed to regional assessments. The emphasis on global assessments reflects a long-standing preference of developing countries, which are always suspicious about the ways national or sub-national data are used. Hence, it can be hoped that regional assessments will be given greater importance in the future, as the regional level of biodiversity governance is certain to gain more prominence (see chapter 3). Other methodological and thematic assessments envisaged in the third objective will represent around 25% of planned expenditures, with capacity-building accounting for 14% of the budget. Capacity-building may therefore take an initial backseat, which may reflect difficulties in developing a model based on a collaborative rather than linear science-policy interface (see infra and chapter 9).

At the regime complex level, assessments may facilitate coordination as well as learning. Though they can be considered as a major tool for improving the science-policy interface, as mentioned by the authors of chapter 4, they may also promote conflicts. Indeed, the assessment-building process makes them instruments that may serve to buttress viewpoints and interests, mobilize groups, build coalitions and justify resource allocation. In fact, judgments about their credibility, legitimacy and relevance may be highly dependent on the political analysis of their impact, rather than the latter on these three conditions.

The challenge for IPBES is significant. Unlike IPCC, IPBES is to serve several conventions and IGOs, some of which had reluctantly supported its creation. The various regimes that form the biodiversity complex are captured by different knowledge communities. As Alice Vadrot (chapter 3) explains

> Throughout the multi-stakeholder meetings, the expectation that IPBES could serve as potential catalyst to create synergies among the six biodiversity-related conventions and the policy issues they address became visible as well. However, the relationship between the CBD and other organizations, namely the World Trade Organization (WTO) and the World Intellectual Property Organization (WIPO), was not at all addressed and hence the potential role of IPBES in this regard neglected.

Thus the question remains as to how an organization such as IPBES – which pools a variety of discourses, interests and visions – could operate and facilitate convergence when it is itself largely the product of only two epistemic communities.

The second dimension of this assumption raises the issue of the management of diverse knowledge systems. One of the main challenges facing IPBES is to create an intersubjective understanding of the contents of biodiversity science despite the different ontologies and epistemologies, competing policy priorities and rival agendas (see chapter 2). Difficulties arise with respect to the role of social sciences, multiscalar biodiversity management, and the integration of local and indigenous knowledge into the analytical framework of positivist Western science. The role and impact of social sciences remain limited (see below). Social scientists (apart from economists) are vastly under-represented in the Multi-disciplinary Expert Panel (MEP). Fewer than five economists and only one anthropologist were among the 62 experts involved in the pollination report (see chapter 11). In addition, greater formal representation will be of little use if social sciences continue to be considered just as adjuncts to natural sciences, mere tools useful for the implementation of solutions to problems defined elsewhere.

Unlike IPCC, though, efforts have been made from the outset to integrate diverse knowledge systems into the work of IPBES through its multi-stakeholder perspective and the insertion of this knowledge into the conceptual framework. Yet the task remains daunting. Concerning indigenous and local knowledge (ILK), for example, it is assumed that so-called "indigenous knowledge" (itself dynamic) can be isolated and that this knowledge is geared toward biodiversity conservation. But what constitutes indigenous knowledge? How can it be isolated from a whole social system of which it is constitutive in order to protect it? How can we build bridges between this type of knowledge and scientific knowledge? Are taboos and other prohibitions really informal systems associated with biodiversity conservation, as many biologists and conservationists believe? Hence the need to take into account the social mechanisms in which indigenous knowledge and practices are embedded (social organization, symbolic systems, modes of transmission, representativeness of various types of knowledge, etc.), something that IPBES is hardly equipped to do (on all these points, see Goedefroit, 2013). This may be because IPBES seems to regard the integration of local and indigenous knowledge more as a problem of participation (ensuring that representatives of local and indigenous organizations are represented in task forces on indigenous knowledge), than as a knowledge-building issue (in which case academics are better placed than representatives of local communities).[1]

The question then remains as to how tensions among various groups of stakeholders will be managed. What criteria will be used to validate whether a given fragment of indigenous knowledge is good or bad for biodiversity? How will the use of traditional knowledge be controlled (see chapter 10)? On what basis will the legitimacy, credibility, and relevance of various types of knowledge be assessed? How will the intermingling of values, interests, and knowledge (a notion that is well documented in the science and technology studies literature, e.g. Jasanoff, 2004) be handled when biologists and other scientists view the relationship between science, values and policy linearly? Scientific debates often mask political struggles (see chapter 3) and science cannot be expected to settle political conflicts – attempts to do so may have the opposite effect. NGOs use knowledge to promote values, indigenous populations to protect rights, firms to identify business opportunities, and states to support policy. A given species may not have the same importance for NGO activists, biologists or local communities.

Reconciling different ontologies and epistemologies may be utopian, if not outright nonsensical, but minimizing conflicts among them is possible. First, it should be recognized that indigenous knowledge does not merely complement scientific knowledge, but instead represents an autonomous and alternate vision of the relationship between society and nature (see chapter 7, as well as objective 2 of the Work Programme). Second, conflicts can be managed through the institutionalization of a specific process. The final diagram of the IPBES conceptual framework marks the recognition of a scientific and a non-scientific vision, but its analysis by Maud Borie and Denis Pesche (chapter 7) reveals that the ecosystem services view is equated with the scientific view, while all other knowledge systems are deemed to represent "Mother Earth".

It has yet to be determined how this convergence within the organization will actually take place. One difficulty lies in politics – how will the organization manage the various power games among parties and stakeholders? Indigenous groups do not consider themselves simple stakeholders, alongside academics, NGOs, industry, scientific and research bodies. Instead, as pointed out in chapter 10, indigenous representatives have stressed that they intend to continue to interact with IPBES in their own capacity as knowledge and rights holders rather than as simple stakeholders (see also chapter 8).

Assumption #2: the problem is a deficient science-policy interface

As recalled in chapter 4, the IPBES work program first aims to "Strengthen the capacity and knowledge foundations of the science-policy interface to implement key functions of the Platform", while its second objective seeks to "Strengthen the science-policy interface on biodiversity and ecosystem services at and across subregional, regional and global levels."

This assumption has two dimensions. The first is the belief, widely held among natural scientists, that policy shortcomings stem from a lack of impact of science on policy. The empirical basis and relevance of this conviction are highly debatable, however. Opponents to the idea of an IPBES questioned it early, with some arguing that science does have an impact (such as the development of specific MEA work programs), and others pointing out that biodiversity loss is more of a normative than a scientific problem. The second dimension is the conviction – stemming from the negative experience of the Global Biodiversity Assessment (see chapters 3 and 4) – that for biodiversity science to have an impact on policy, policy makers have to be associated with its development and be able to convey their research needs.

The main implication for governance of the Platform has been to try to set up and implement a collaborative model of science-policy relations, as a counterpoint to the prevailing linear model (Koetz, Farrell, & Bridgewater, 2011; Van den Hove, 2007), as detailed in various chapters. Yet, translating this approach into a working model able to guide the work of IPBES (and not only its formal participation structure) is hardly straightforward. As chapter 2 points out in reference to the first assessment on pollinators, the MEP has yet to fully integrate the non-linear science-policy relationship model that IPBES is supposed to embody. For example, instead of starting by looking at how various stakeholders define the problem, scientists are relied on to identify trends the reversal of which is defined a priori as the issue that must be addressed. Indeed, as the authors of chapter 11 point out, this choice of topic did not result from strong demand from governments and stakeholders. They suggest that the very procedures that IPBES has put in place for conducting this assessment have served to attenuate the controversy on pesticides rather than shedding light on the issue in order to map out the contours of appropriate policy responses. In that regard, they also raise an intriguing point as to a potential trade-off between legitimacy and the expression of socio-technical controversies. Capacity-building is a related concern and essential for bringing science to policymakers insofar as governments tend to listen to their own scientists rather than foreign scientists, who are often suspected of defining problems and promoting solutions that do not reflect the perspectives and interests of the host country.

Assumption #3: the Platform's legitimacy and effectiveness hinge on its ability to include all stakeholders in the development of policy-relevant knowledge

Eager to build up the science-policy interface, the IPBES founders chose to remain strictly intergovernmental, thus eschewing a dual or tripartite public-private model. IPBES nevertheless aims to mobilize a large array of stakeholders to support the implementation of its work program. In turn, stakeholders and observers have staked a claim in favor of their active participation in the IPBES process. In many ways, the resulting stakeholder engagement strategy is an institutional innovation that has yet to be seen in other science-policy interfaces (see chapters 5 and 8). In contrast with IPCC, the work program elaboration process allows for variation between evaluation cycles, and the MEP relies on requests, suggestions and inputs from members, observers and stakeholders before the work program is adopted in the Plenary. Yet, priority is given to requests from states, which in practice means that suggestions by other stakeholders must be supported by one or several governments before progressing further.

The participation of scientists from developing countries is also a matter of concern. As pointed out in chapter 6, one shortcoming of IPCC is related to links between its composition and its legitimacy, notably with respect to the under-representation of scientists from developing countries and generally of social scientists (with the exception of economists). The MEP seeks to overcome that geographic limitation by being structured on a regional basis. That experts are directly involved as coordinating lead authors (CLAs) or lead authors (LAs) in the assessments is a clear reflection of this openness strategy. In this regard, resource constraints may impede that process unless governments step in. Chapter 6, for example, points out that in the case of IPCC (and IPBES is not different), CLAs need external support, in terms of human resources, to carry out their demanding task. In this context, the IPBES budgetary contributions will have to be significantly diversified. As mentioned in chapter 4, there is indeed a clear imbalance in financial contributions, with 65% of the budget provided by only two countries. Indeed, redressing this situation has become the main objective of Robert Watson, President of IPBES.

As Guillaume Futhazar explains (chapter 5), the reconciliation of the desire of states to keep control of the nomination process with the need to broaden the participation of stakeholders "gave rise to a convoluted solution", with MEP members, nominated by member states according to a geographical balance, in turn selecting experts in charge of drafting each work program report from a list of experts nominated by both governments and relevant institutions. However, the selection has to respect a very precise balance that calls for a quota of 80% of government-nominated experts and 20% of experts from relevant institutions.

The need and willingness to include a variety of stakeholders raises a number of thorny issues, some of which came up during the first assessment on pollinators (chapter 11). Beyond the scope of participation, one issue concerns the dilemmas and difficulties associated with selecting experts from among a diverse range of legitimate stakeholders whose unequal capacities and power may either facilitate or hinder attempts by the Secretariat to influence their role and contributions to the process (see chapter 8).

Another challenge is how to manage the different visions and priorities of academics (concerned with the advancement and use of knowledge), local and indigenous populations (whose positions reflect political claims), and NGOs (more concerned with the promotion of specific values). As the authors of chapter 8 remind us, one of the major challenges in global governance research is to gain further insight into the dynamics of power and influence among this diversity of stakeholders. As mentioned earlier, ILK representatives see their status as stakeholders as "a step backwards and a way of minimizing their ability to produce knowledge considered as legitimate." More generally, "The idea of a group of stakeholders, promoted by IPBES, is accompanied by a leadership/participation model that tends to either enrol the stakeholders involved or conversely, lead them to exclude themselves from the stakeholders' dynamic. The most contentious categories, as heterogeneous as they are, "might then prefer to rule themselves out" (chapter 8).

Finally, participation can be instrumentalized. For example, Duperray and colleagues (chapter 11) point out that, "the process used to select scientific experts was also particularly political insofar as the scientific and policy positions of the experts were taken into account in their recruitment", thus minimizing the controversial aspects of the issue. Indeed, participation can be thought of as an end in itself (to legitimize outputs), or as a means to an end (to improve decision making and outcomes). It can be used to impose a particular viewpoint (through the selection of participants) and thus steer policy dynamics in a preferred direction. It is not participation that shapes politics but the other way around.

Conclusion

An indirect product of IPCC, IPBES also seeks to develop a new assessment model and science-policy interface. In so doing, as Guillaume Futhazar (chapter 5) suggests, IPBES will certainly develop in a very different direction from that of IPCC. Several chapters examine how IPBES – drawing from the experience of IPCC and other assessments – has attempted to implement the various principles and objectives it pursues, such as participation, capacity-building, or the inclusion of traditional knowledge.

As Alice Vadrot points out, institutions are the product of power relations and interests. They promote or maintain unequal power relationships (between groups, types of knowledge, disciplines, stakeholders, parties, etc.) through a dominant discourse that itself defines what is considered relevant knowledge. Conflicts regarding biodiversity and the science-policy interface configuration mask conflicts over power and resource distributions.

In this context, according to the authors of chapter 11, the pursuit of these objectives will not be easy. Although "IPBES became the ideal arena for coordinating an assessment of the stakes involved in several regime complexes (food security, agriculture, biodiversity, trade, etc.)", its value is still virtual as its procedures seem to be hampering progress to that end. How to balance the various expectations from other members of the complex will be delicate. IPBES procedures have been put in place (see chapter 3), however, in practice, the first assessment on pollinators addressed a CBD concern and was largely framed along CBD-oriented lines. Avoiding IPBES becoming the equivalent of IPCC for UNFCCC thus remains a challenge.

IPBES represents a disruptive innovation in a complex system of international biodiversity governance as it tries to define a new niche for itself, thereby forcing other system units to adapt and change their relationships. We cannot assume that everything else will remain constant and that IPBES will only help coordinate the scientific functions of various organizations, while achieving consensus about what is considered relevant, legitimate, and credible knowledge, thereby strengthening the science-policy interface.

Ultimately, the key issue is learning, which can be approached at three levels of increasing depth: (i) simply as institutionalized error correction (learning by experience); (ii) as the capacity to integrate new knowledge into behaviour, or to respond more quickly to new demands and new information (Hedberg, 1981), which is crucial as the policy environment becomes more uncertain; and (iii) according to E. Haas (1990), as a fundamental change in the nature of the consensual knowledge of actors, i.e. a change in the norms built into theory-in-use accompanied by a change in the organization's model of the world (Haas, 1990). So to what extent can IPBES play a role in complex learning, where agents gradually develop shared interpretations of the problem, and where these interpretations come to define their identity, the shape of the institutions, and the possibility of cooperation? It is thus essential to study the internal (relationships among stakeholders and various knowledge systems, capacity to put forth alternative ideas, leadership, the existence of learning agents) and external (stable external dominant coalition of states and NGOs) conditions that will affect the capacity of

IPBES to put in place the structure and procedures that will enable it to gain the most from its focus on capacity and network building.

Note

1 The one anthropologist sitting on the MEP, however, is an academic.

References

Goedefroit, S. 2013. "*La nature est culture.*" Paper presented at the Congrès de l'ACFAS; Forum EDS sur "Développement durable et biodiversité: le rôle des universitaires", Université Laval (Québec), May 8, 2013.

Haas, E. B. 1990. *When Knowledge is Power: Three Models of Change in International Organizations.* Berkeley: University of California Press.

Hedberg, B. 1981. "How Organizations Learn and Unlearn," in *Handbook of Organizational Design*, P.C. Nystrom and W.H. Starbuck, eds. Oxford: Oxford University Press, pp. 3–27.

Jasanoff, S. (ed.) 2004. *States of Knowledge: The Co-production of Science and Social Order.* London: Routledge.

Koetz, T., Farrell, K. N., & Bridgewater, P. 2011. "Building Better Science-policy Interfaces for International Environmental Governance: Assessing Potential within the Intergovernmental Platform for Biodiversity and Ecosystem Services." *International Environmental Agreements: Politics, Law and Economics* 12(1): 1–21.

Van den Hove, S. 2007. "A Rationale for Science–Policy Interfaces." *Futures*, 39, 807–826.

Index

Page numbers in italics refer to figures. Pages numbers in bold refer to tables.